Natures of Data

PHILIPP FISCHER, GABRIELE GRAMELSBERGER,
CHRISTOPH HOFFMANN, HANS HOFMANN,
HANS-JÖRG RHEINBERGER, HANNES RICKLI

Natures of Data

**A Discussion between Biology, History
and Philosophy of Science and Art**

DIAPHANES

Table of Contents

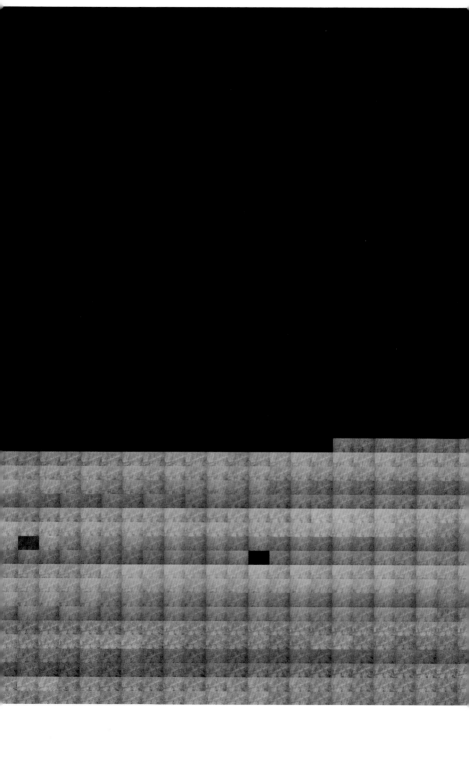

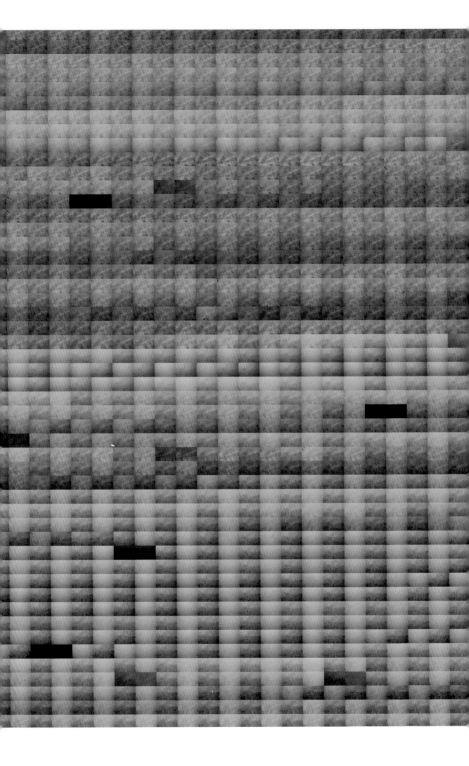

Introduction

The nature of data can mean: What kind of nature takes shape in captured data? It can also mean: What is the nature of data, what characterizes them? And finally, under the nature of data one can also imagine an environment, a biotope, in which data flourish and endure. The title of this book incorporates all three of these meanings—and thus also expresses that nature and data are interrelated in multifarious ways. No particular occasion is needed to reflect upon scientific data in this manner. In the modern sciences, data, even though the pertinent meaning of the word did not become established until the twentieth century,[1] constitute the undisputed point of reference of all knowledge. Nevertheless, since the 1990s the situation has changed once more—in some areas of the sciences much earlier, later in others. From this point on, data increasingly make their appearance digitally, packed in files. From this point on, calculations and graphic renderings become more and more a matter for programs. From this point on, one device dominates scientific institutions: the computer. Everywhere, writes philosopher Michel Serres, one encounters "the same picture: people sitting at computer screens, pounding on their keyboards. They can no longer be distinguished in this dance of bodies. Be they scholars, a species that is slowly disappearing, be they biologists or astrophysicists, chemists or topologists, they all struggle at the same machine."[2] Although this observation is somewhat exag-

---

1 See Daniel Rosenberg, "Data before the Fact," in *"Raw data" is an Oxymoron*, ed. Lisa Gitelman (Cambridge, MA: MIT Press, 2013): pp. 15–40.
2 Michel Serres, preface to *Le Trésor: dictionnaire des sciences*, ed. Michel Serres and Nayla Farouki (Paris: Flammarion, 1997): pp. VII–XXXIX, XI.

gerated, Serres was right about the general tendency. And a second thing about his description provides food for thought: Not only do the differences between the scientific professions disappear on the computer, it seems as if what they do there can no longer be traced. Yet, this is not the case: the actions through which science is performed on the computer are by no means beyond control and observation. But new kinds of awareness must be developed in order to do so.

The vanishing traceability and the increasing dematerialization of research through its relocation to the digital realm were the premise of the *Computer Signals* project, from which this publication emerged. This premise, as suggested, was not corroborated. The longer and more deeply we concerned ourselves with the circumstances of these research worlds organized around computers, programs and data, the more clearly three aspects came to the fore: The diversity of the work to be performed; the playful, experimental element of handling data and programs; and the powerful infrastructures in which scientists pursuing their research are embedded. Each of these aspects leads us in a different direction. Taking an interest in data work means being aware of how laborious this work is, which uncertainties accompany it, and what new kinds of expertise it requires. Taking an interest in the "dance with the data" (as a participant in the following conversation called it), leads away from the idea of somehow automating the research process; on the contrary, in the digital realm research is no less *bricolage*, testing and scrabbling than it used to be, it just takes place in a different environment. With the infrastructures, finally, once again, what comes forth is not exactly new. Research has always been dependent on a material framework that makes it possible: even a kitchen table is an infrastructure. What is new is how far reaching these infrastructures are—and how delicate.

Even though the interaction among data, computers and programs is quite mundane and goes on in full sight of the world, its circumstances and consequences are not yet fully understood. In all of its meanings outlined above, the nature of data could have been structured in a new way—or not. This is precisely what this book is about. Rather than providing answers, it undertakes test drillings. While digitization marks all sciences, we limit our discussion to biology, for which the processes in question certainly do play a central role. And we undertake these probes not in the form of completed essays, but in a form appropriate to the project and its history: as a conversation. Calling *Computer Signals* a project is formally correct; it includes everything a project needs. But in fact, *Computer Signals* is an ongoing meeting between biology, art, philosophy of science and history of science. We have been talking with each other regularly for over five years now (some of us even longer), discussing biological and artistic projects, observing, learning, and occupying ourselves with these new forms of scientific work. Generally one would speak of an interdisciplinary context, yet the different origins of the participants matter only insofar as they force us to come to an understanding with each other, and that means, not least, to express our own premises and requirements.

The discussion group was formed in Hannes Rickli's first project, *Surplus: Videograms of Experimentation* (2007-2009), which reflected artistically and theoretically on a bundle of analog video recordings that were unsystematically compiled from laboratory contexts starting in the 1990s, presenting them as visual and simultaneously physical traces of the production of scientific facts.[3] His interest was focused on the use of technical visual media in the process of research. If analog, unedited image remnants allowed the interplay of media, spatial, human and animal actors as a resistant network of relationships to be read retrospectively, in the transition to digital techniques it became apparent that the processes

---

3   "Überschuss: Videogramme des Experimentierens," https://www.zhdk.ch/forschungs-projekt/426832. See also Hannes Rickli, ed., *Videograms: The Pictorial Worlds of Biological Experimentation* (Zurich: Scheidegger & Spiess, 2011).

of research were withdrawing ever more from the human senses, thus making direct observation difficult and sometimes impossible. The decisions about the scientific relevance of recorded signals and their evaluation are left to algorithms. Anything that is not automatically deleted from digital data is consistent with the program and thus tends to be uninteresting from the artistic perspective.

The work of biology consists in isolating phenomena in confusing environments, applying complex methods of abstraction with digital programs in black boxes in order to cleanse them of the effects of particular and singular elements, and allowing them to be transported as distinct numbers or formulas. The work of art runs in the other direction: It is interested in the material conditions of this abstraction, reconstructing the concrete circumstances in which biological data are collected, distributed and calculated. As such, it shifts the focus to the time spent, the concrete spaces and lighting conditions, the scientists' gestures, and, last but not least, the animals at the focus of the biological knowledge interest.

In the projects *Computer Signals I* and *Computer Signals II* (2012-2015; 2017–2020), the artist Valentina Vuksic developed a special experimental system that records the processes in the infrastructures of the electronic data work by Hans Hofmann at The University of Texas at Austin and that of Philipp Fischer on the islands of Heligoland and Spitsbergen, who records time-based audio files. Electromagnetic pickups and contact microphones placed parasitically in the measuring instruments and computers realize a "technical participant observation" of the devices' operation in their environment, which "listens" to how they function, fluctuate, falter and interrupt.[4] They register the variations in energy consumption, the electromagnetic oscillations at rest and when a camera

---

4   See Shintaro Miyazaki, *Algorhythmisiert: Eine Medienarchäologie digitaler Signale und (un)erhörter Zeiteffekte* (Berlin: Kulturverlag Kadmos, 2013).

shutter is released, or the normally inaudible rhythmic clicks of a computer processor working through algorithms. The synchronously recorded mechanical vibrations of the devices establish connections to the ocean currents in the Kongsfjord off the coast of Spitsbergen, or to the high-revving pumps of a cooling system keeping the *Stampede* supercomputer running in the hot Texas climate. The art project first had to learn how to deal with the amounts of data accrued daily in the project's own server at the Zurich University of the Arts. The difficulties that arise point out the social and communicative, electrical and technical imponderables that characterize data work, in art as well as in biology. The art project discusses the material obtained and the experiences of dealing with them in exhibitions (among others, Schering Stiftung, Berlin, 2013; Kunstraum Walcheturm, Zurich, 2020), at international conferences, and in this publication.

For the subsequent conversation we met high above Lake Lucerne on Rigi Kulm in early September 2016. Our intention was to bring together the observations, considerations and impulses of the previous years into a loosely arranged form. For this we set up four blocks of conversations: First came Data, followed by Software and Infrastructures; we finished by following a case study into the world of *in silico* research. The transcripts have been reworked and abridged, but we preserved as much of the original conversational character as possible. We wander through the topics tentatively, keeping just a few keywords in mind, sometimes digressing, sometimes coming back to what someone said earlier. We now briefly sketch our approach to the four sections: Data (Christoph Hoffmann), Software (Gabriele Gramelsberger), Infrastructures (Hannes Rickli) and *in silico* (Hans Hofmann). The book closes with an epilog by Hans-Jörg Rheinberger, who recalls the origins and developments of our discussions since the mid-1990s.

Data

A good ten years ago it looked as if the sciences, and the life sciences most of all, were at the threshold to a new age. The corresponding keywords were: *Big data, data-driven research, Fourth Paradigm.* Henceforth, statistical analyses of great amounts of data would bring new insights almost by themselves; without any previously formulated hypotheses and without support through observations and experiments. This vision has since lost much of its glamor. Philosophers of science objected that scientists do not make advances without relying on background knowledge, indeed, hypotheses in the congealed form of theories and concepts come into play in the very definition of datasets, what is paid attention to, and in the preceding generation of the data used.[5] A look at the history of sciences also makes clear that not much terminological contortion is necessary to speak of the natural historical collections since the 18th century as data-driven research.[6] Such reservations were reinforced by another element: A multitude of studies pointed out that it is by no means a trivial activity to assemble data from different sources.

Data have a history that consists of the circumstances of their generation—the technologies deployed and the particular research questions, which must be borne in mind when they are used in a new context.[7] This circumstance hinders

---

5   Sabina Leonelli, "Introduction: Making Sense of Data-Driven Research in the Biological and Biomedical Sciences," in *Studies in History and Philosophy of Biological and Biomedical Sciences* 43 (2012): pp. 1–3; Werner Callebaut, "Scientific Perspectivism: A Philosopher of Science's Response to the Challenge of Big Data Biology," in *Studies in History and Philosophy of Biological and Biomedical Sciences* 43 (2012): pp. 69–80.

6   David Sepkoski, "Towards 'A Natural History of Data': Evolving Practices and Epistemologies of Data in Paleontology, 1800–2000," in *Journal of the History of Biology* 46 (2013): pp. 401–444; Staffan Müller-Wille, "Names and Numbers: 'Data' in Classical Natural History, 1758–1859," in *Osiris* 32 (2017): pp. 109–128; Bruno J. Strasser, *Collecting Experiments: Making Big Data Biology* (Chicago: University of Chicago Press, 2019).

7   Geoffrey Bowker, *Memory Practices in the Sciences* (Cambridge, MA: MIT Press, 2005): chap. 4.

the circulation of data; leads to questions about standardization; raises awareness of the metadata included in datasets as a critical point for their storage, dissemination and renewed utilization; and directs interest toward ontologies and databases as tools that not only ease circulation, but also channel access and questions.[8] "Data travel," in other words, is complex and time consuming, and inevitably accompanied by conflict—"data friction."[9] Yet formal hurdles are not the only thing hindering the circulation of data. Researchers who rely on third-party data are faced with the problem of having to understand the properties of these datasets and to assess their quality for the intended use. Personal experience, knowledge of traps, tests, and informal inquiries can help.[10] What is clear, however, is that data cannot simply be used without accounting for the context of their generation; instead, a number of elaborate operations are needed, combined with a healthy portion of trust. And yet another point must be mentioned in this connection: Data have a value; their generation makes work. Sharing data is therefore not a matter of course. While state regulations and guidelines from funding institutions increasingly stipulate that data sharing is mandatory, stor-

8   Bowker, *Memory Practices in the Sciences*: chap. 3; Christine Hine, "Databases as Scientific Instruments and Their Role in the Ordering of Scientific Work," in *Social Studies of Science* 36 (2006): pp. 269–298; Sabina Leonelli, "Packaging Small Facts for Re-use: Databases in Model Organism Biology," in *How Well Do Facts Travel: The Dissemination of Reliable Knowledge*, ed. Peter Howlett and Mary S. Morgan (Cambridge: Cambridge University Press, 2011): pp. 325–348.

9   On the circumstances of "data travel," see Sabina Leonelli, *Data-Centric Biology: A Philosophical Study* (Chicago: University of Chicago Press, 2016): chap. 1; on "data friction," see Paul N. Edwards, *A Vast Machine: Computer Models, Climate Data, and the Politics of Global Warming* (Cambridge, MA: MIT Press, 2010): chap. 5.

10 Ann S. Zimmerman, "New Knowledge from Old Data: The Role of Standards in the Sharing and Reuse of Ecological Data," in *Science, Technology & Human Values* 33 (2008): pp. 631–652; Paul N. Edwards, Matthew S. Mayernik, Archer L. Batcheller, Geoffrey C. Bowker, and Christine L. Borgman, "Science Friction: Data, Metadata, and Collaboration," in *Social Studies of Science* 41 (2011): pp. 667–690; Goetz Hoeppe, "Working Data Together: The Accountability and Reflexivity of Digital Astronomical Practices," in *Social Studies of Science* 44 (2014): pp. 243–270.

ing data in publicly accessible repositories (in this case public generally means accessible to members of a community) does not necessarily mean that they will be used. Which data come into circulation depends not only on whether they fulfill the standards of a certain database. What matters most is whether they meet the special demands of the users.

It is no coincidence that these aspects are frequently discussed with in regard to in the life sciences. Since the sequencing projects of the 1990s, and even more so in the -omics world of today, the generation and management of large amounts of data, supercomputers, programmed analysis tools, and the close link to information technology are part and parcel of everyday science.[11] This development has also long arrived in more traditional areas as well: ecology, behavioral science, systematics—algorithms, databases and computational methods are being integrated in the research process everywhere. The life sciences are particularly interesting, however, also because of the specificity of their data. They are often local, bound to certain locations, sometimes demand long time series, and can be extremely heterogenous in one single project. Moreover, variability is an essential characteristic of biological objects. These data are marked by peculiarities on the level of the individual, species and population, an aspect that must be taken into consideration when they are compared or transferred into other contexts. After all, the life sciences comprise a multitude of disciplinary cultures, each of which has its own research approach and concepts, which flow into the generation, processing and description of data. This is why data work is such a frequent focus in the life sciences. Its circumstances make it much richer than in most other sciences.

---

11   Hallam Stevens, *Life Out of Sequence: A Data-Driven History of Bioinformatics* (Chicago: University of Chicago Press, 2013); Leonelli, *Data-Centric Biology*; Strasser, *Collecting Experiments*, chap. 4–6. However, development began much earlier in the early 1960s, see Joseph November, *Biomedical Computing: Digitizing Life in the United States* (Baltimore: Johns Hopkins University Press, 2012).

Undoubtedly, biological research has changed in recent decades. Besides the wet lab there is now a dry lab, often in the same room. Sensors stream measurement data that had to be recorded by hand in the old days. Audio and video recordings are controlled by algorithms, and to some degree, are already analyzed using machine learning approaches. This results in new divisions of labor, the organization of research processes changes, required new skills and expertise have to be integrated, training programs expanded, and the distribution of credits and merits adjusted. A closer look reveals a second meaning of the term data-driven research. Taking care of the data, data maintenance, archiving, which were once secondary activities, are becoming ever more elaborate. Data are not only valuable; increasingly, they are conceived as an expensively generated resource that must be used well. What is most striking, however, is how data, traditionally understood as products of observations and experiments, have themselves become the object of observations and experiments, and how research is thus shifting into the data space.

Software

In his article "The Teaching of Concrete Mathematics" from the year 1958, the statistician John W. Tukey first defined the term "software." Against the backdrop of the early development of computers and the strengthening of applied mathematics, he advocated that "today the 'software' comprising the carefully planned interpretive routines, compilers, and other aspects of automative programming are at least as important to the modern electronic calculator as its 'hardware' of tubes, transistors, wires, tapes and the like."[12] This was hardly a matter of course in the 1950s. Back then

---

12   John W. Tukey, "The Teaching of Concrete Mathematics," *The American Mathematical Monthly* 65 (1958): pp. 1–9, 2.

computers still had names, like *DERA* (Darmstädter Elektronischer Rechenautomat), *ERMETH* (Elektronische Rechenmaschine der ETH Zurich), *Electrologica X1* (Mathematisch Centrum Amsterdam), and the 123 systems of the IBM 704 series, which were the products of heroic engineering skills. What is particularly interesting about this last computer is that, in October 1956, it was delivered with the first automatic coding system.[13] This automatic coding system was Fortran (Formula Translation), which today is considered to be the first high-level programming language. Fortran automated the configuration of computers through machine commands. Instead of machine commands, programmers were to use "a concise, fairly natural mathematical language" to program software for "mechanical coding."[14] Mathematics was chosen to function as the natural language because most users of the IBM 704 series were engineers and natural scientists, who delegated the calculations of their mathematical models to automatic computing machines.[15] Tukey's 1958 definition of software referred back to this tradition.

Today scientific research takes place ever more frequently in software—algorithms, programs and the corresponding data. The calculation of models has been augmented by a plethora of other software applications and software has become the preeminent research instrument. It can be used to

---

13   John W. Backus et al., *The FORTRAN Automatic Coding System for the IBM 704 EDPM: Programmer's Reference Manual* (New York: Applied Science Division and Programming Research Department: International Business Machines Corporation, 1956).

14   "At that time [1954], most programmers wrote symbolic machine instructions exclusively ... they firmly believed that any mechanical coding method would fail to apply the versatile ingenuity which each programmer felt he possessed and constantly needed in his work." John W. Backus and William P. Heising, "FORTRAN," *IEEE Transactions in Electronic Computing* 13 (1964): pp. 382–385, 382; see also Gabriele Gramelsberger, "Story Telling with Code," in *Code: Zwischen Operation und Narration*, ed. Andrea Gleininger and Georg Vrachliotis (Basel: Birkhäuser, 2010): pp. 29–40.

15   John W. Backus and Harlan Herrick, "IBM 701 Speedcoding and other automatic programming systems," *Proceedings of the Symposium on Automatic Programming for Digital Computer* (Washington, DC: Office of Naval Research, Department of the Navy, 1954): pp. 106–113, 112.

digitize and control experimental systems and measurement instruments, to generate and process tremendous amounts of data, and to simulate even elaborate models and extrapolate them into the future. It almost seems as if natural scientists today research their software products more than nature itself.[16] The example of biological behavioral science shows how, thanks to software, the various bodies of knowledge and knowledge practices migrate into computers. In the case of digitized fish observation this entailed many changes for research; above all, the automation of the observer. While researchers used to perform the time-consuming work of viewing the video material of fish observations, now software analyzes the digital recordings of the fish and their behavior in a matter of seconds—for instance, the production of noises to research fish sounds. In this case, analyzing means automatically registering their behavior in the video and audio recordings. Thus the software decides, according to coded instructions by the programmer, whether anything interesting for behavioral biology happens, which it then saves as events and renders in analysis curves. The image and audio recordings themselves are deleted. While researchers previously observed the animals via videos, now the software observes the digital recordings of the animals. The researcher, in turn, observes the software, or at least its analysis curves, and, if need be, the remainder of observational materials which the software was not able to categorize. While observation of the video recordings previously led to quite arbitrary speculation about the fish that produced a sound (assumption: the fish that moved), today the source of the sound can be located quite accurately using three hydrophones and suitable triangulation software. Thanks to the software, an assumption becomes a fact. Moreover, the localization also allows considerably more precise analysis of fish behavior. In order to

16 See Gabriele Gramelsberger, *Computerexperimente: Zum Wandel der Wissenschaft im Zeitalter des Computers* (Bielefeld: Transcript, 2010).

investigate this in greater detail, even more digitalization and analysis algorithms are needed, and so on. An endless chain of processing ever new data on fish;[17] but also constant transformations of research practice, which keep an accelerating digitalization cycle going.

It is not possible to even estimate how much software is produced in science each year. But what can be quantified are the increasingly comprehensive cyberinfrastructures for the development and publication of scientific software. Scientific software is supposed to be freely available, and these days it is usually developed collaboratively using GitLab or Jupyter. The result of such collaboration can be cited by a digital object identifier (DOI) as a "text product."[18] Databases like runmycode.org and execandshare.org publish software products that belong to scientific articles. Projects like *CodeMeta* provide meta descriptions of scientific software in JSON (JavaScript Object Notation) or XML (Extensible Markup Language). All of this is required in order to make software-based research reproducible. While the methodological and the epistemic value of research has been guaranteed by transparency, and transparency through reproducibility since the early modern age, the multitude of methods and software products today, not to mention the vast amounts of data, makes it ever more difficult to reproduce results. On the one hand, there is the development that, "Within science, reproducibility is threatened, among other things, by new tools, technologies, and big data."[19] On the other hand, thanks to software and the

17 Philipp Fischer, "Datenströme in der marinen Verhaltensökologie: Eine Herausforderung an die moderne Wissenschaft," paper presented at the conference *Fragile Daten*, Berlin-Brandenburgische Akademie der Wissenschaften, Berlin, March 1–2, 2013, https://epic.awi.de/id/eprint/32555/.

18 Nick Barnes, "Publish Your Computer Code: It is Good Enough," *Nature* 467 (2010): p. 753.

19 Harald Atmanspacher and Sabine Maasen, "Introduction," in *Reproducibility: Principles, Problems, Practices, and Prospects*, ed. Harald Atmanspacher and Sabine Maasen (Hoboken, NJ: John Wiley & Sons, 2016): pp. 1–8, 1.

internet, it has never been easier to disclose scientific methods, software products and results and thus make them reproducible.[20]

Infrastructures

Data must always be related to the technical apparatuses that collect, calculate and distribute them. These are dependent on technological services like electricity, computer networks, satellites and submarine cables, pipelines, ready-made algorithms, and under certain circumstances, also bioactivity: for example worms, algae, icebergs and storms. Apparatus and infrastructures make research possible, while at the same time setting its limits. They usually operate below the user's perception and are "considered to be a hidden substrate— the binding medium or current between objects of positive consequence, shape, and law," in the words of city planner Keller Easterling.[21] They become visible only when they fail or break down, and then, by deviating from their intended functionality, manifest chaotic dynamics.[22] In such cases, the "technical objects," as Hans-Jörg Rheinberger calls them, can be observed in their unintended autonomous activities.[23]

---

20 "Now scientists routinely share preprints, published papers, and other forms of traditional scientific knowledge transmission mechanisms, but they also share entirely new forms such as datasets, code, high resolution images, software designed to entail the manipulation of results by others, links and lists of related works. This facility can make the black boxes significantly easier to open." Victoria Stodden, "The Scientific Method in Practice: Reproducibility in the Computational Sciences," *MIT Sloan Research Paper No. 4773-10* (Cambridge, MA: MIT Sloan School of Management, 2010): p. 28.

21 Keller Easterling, *Extrastatecraft: The Power of Infrastructure Space* (London: Verso, 2014): p. 11.

22 See Jane Bennet, "The Blackout," in *Vibrant Matter: A Political Ecology of Things* (Durham: Duke University Press, 2010): pp. 24–28.

23 "It is through them (the technical objects; HR) that the objects of investigation become entrenched and articulate themselves in a wider field of epistemic practices and material cultures, including instruments, inscription devices, model organisms, and the floating theorems or boundary concepts attached to them. ... The experimental

In order to control them in a given research project, the unique characteristics of their behavior during failure must be studied. Rheinberger describes this shift in cognitive interest as the shift from the "epistemic thing" to the "technical objects," in which infrastructure and its surroundings themselves become the object of study. In this process the interests swing back and forth like a pendulum between the one thing and the other, with the infrastructures binding the bulk of the resources for undertaking research.

What is interesting about infrastructures from the artistic perspective is their multifold play between inconspicuousness, autonomous activity and massive materiality. Do they really play only a serving role? Or do they instead determine what we can know, by organizing the mesh of relationships between human and nonhuman actors in the cognitive process? Michel Serres might perhaps designate infrastructures as "quasi objects," as something that "approaches zero," which remain invisible in use, even though they structure the relations between objects and people in their environments, although one never really knows whether they are actually objects or subjects.[24]

Terms like assemblage, collective, and network are used to describe the confluences in which technical, "natural" and social elements cooperate and conflict with each other.[25] In connection with the life sciences, the following interesting constellation emerges: nature is the object of research *and* simultaneously the medium of its exploration. Nature provides

conditions 'contain' the scientific objects in the double sense of this expression: they embed them, and through that very embracement, they restrict and constrain them." Hans-Jörg Rheinberger, *Towards a History of Epistemic Things: Synthesizing Proteins in the Test Tube* (Stanford, CA: Stanford University Press, 1997): p. 29.

24 See Michel Serres, "Theory of the Quasi-Object," in *The Parasite*, trans. Lawrence R. Schehr (Baltimore: Johns Hopkins University Press, 1982): pp. 224–234.

25 See Bennet, *Vibrant Matter*: pp. 20–38; Serres, "Theory of the Quasi-Object"; Bruno Latour, *Reassembling the Social: An Introduction to Actor-Network-Theory* (Oxford: Oxford University Press, 2005).

the raw materials for energy production and is the carrier of the terrestrial, submarine and orbiting distribution channels for electricity and information. These massive structures, in turn, are dependent on a lower layer, generally on fossil fuels, which deliver the energy to manufacture computers, cool server farms, and transport material and maintenance staff to remote regions. In this sense, nature itself becomes the "ultimate infrastructure."[26]

While technical infrastructures and energy supply are developed in a process of societal negotiation, and are in a state of constant adaptation, the energetic activities of nature are not negotiable. In the form of algae and worm growth, storms and icebergs, they contribute continuously or abruptly to the research process at the submarine station *RemOs1* in the fjord near Ny-Ålesund in Spitsbergen, causing turbidity in the optical system, and electrical short circuits through corrosion or collision. Here nature reveals itself to be a fellow player with agency so strong that it can hardly be brought under control, which keeps the research enterprise in a precarious state. The infrastructures at the University of Texas at Austin, in contrast, have a much more stable basis. Yet the deployment of large-scale technologies like supercomputers and oil fields that belong to the university raises new questions. Including: whether the debate about the ecology, economy and politics of fossil energy resources in the age of climate change should be extended to address epistemic aspects, to illuminate how knowledge horizons are enabled and restricted under the conditions of technology-based, data-driven research. The foundation for such a debate would be for scientists, and especially scientific communication, to report not only on research results, but also on the processes of producing data and objects.

---

26 Nicole Starosielski, "Fixed Flow: Undersea Cables as Media Infrastructures," in *Signal Traffic: Critical Studies of Media Infrastructures*, ed. Lisa Parks and Nicole Starosielski (Urbana, IL: University of Illinois, 2015): pp. 53–70, 54.

In silico

Across the life sciences, the availability of big data has begun to transform how scientists ask questions and carry out their research. For example, many petabases of RNA and DNA sequence data are now available in public repositories, and the influx of new data is ever accelerating. In fact, it has been projected that by 2025 the demands on data storage and processing created by this exponential increase in -omics data (i.e., data resulting from the genome-scale characterization and quantification of biological molecules that underlie the structure and function of cells and organisms) will eclipse other data-intensive areas such as astronomy and social media.[27] The computational analysis, modeling, and experimentation that takes advantage of large and diverse biological datasets is sometimes referred to as *in silico* biology, an allusion to the commonly used Latin phrases *in vivo*, *in vitro*, and *in situ*. *In silico* biology utilizes the vast amounts of biological information available and applies advanced algorithms, models, and simulations to advance scientific understanding. The results of these analyses lead to predictions that can then be tested experimentally or serve as a benchmark for future physical experimentation.

This wealth of data creates tremendous opportunities for re-using publicly available data to gain new insights into long-standing biological questions by integrating -omics data from multiple approaches and studies into innovative meta-analyses that utilize sophisticated computational, bioinformatic, and statistical tools.[28] The potential and limitations of this new big data biology have been discussed for some time,[29] and

27 Zachary D. Stephens, Skylar Y. Lee, Faraz Faghri, Roy H. Campbell, Chengxiang Zhai, Miles J. Efron, Ravishankar Iyer, Michael C. Schatz, Saurabh Sinha, and Gene E. Robinson, "Big Data: Astronomical or Genomical?," *PLoS Biology* 13, no. 7 (2015): pp. 1–11.

28 Sabina Leonelli, "Philosophy of Biology: The Challenges of Big Data Biology," *eLife* 8 (April 2019), doi: 10.7554/eLife.4738.

29 Priyanka Bhandary, Arun S. Seetharam, Zebulun W. Arendsee, Manhoi Hur, and Eve

data management plans have become a standard requirement by many funding agencies. Yet there is currently no generally agreed upon convention on where and how to store this data, and what kind of metadata (information about experimental design, biological samples, protocols, etc.) should accompany raw and processed data. Similarly, the acquisition, storage, distribution, and analysis of these datasets relies on an advanced high-performance computing (HPC) infrastructure, although many technical and conceptual challenges have to be addressed.[30] Finally, the advent of big data biology also raises important questions about the nature of "data" itself, about what constitutes a good dataset, and whether the knowledge gained from such data can be as reliable as traditional knowledge.

An example of one such *in silico* project was the focus of the last part of our conversation. The point of departure for this discussion was a phenomenon referred to by biologists as the developmental hourglass.[31] In the nineteenth century, Karl Ernst von Baer noticed that embryos of a given lineage exhibit the most morphological similarity across species during mid-embryogenesis (a developmental pattern resembling an hourglass), a phenomenon that Klaus Sander designated the "phylotypic period." Even though this phenomenon was confirmed in the early twentieth century, testable hypotheses as to the biological basis of the phylotypic period were largely lacking until about 25 years ago, when two of the founders of the modern field of evolutionary development (evo-devo),

Syrkin Wurtele, "Raising Orphans from a Metadata Morass: A Researcher's Guide to Re-use of Public 'omics Data," *Plant Sci* 267 (2018): pp. 32–47; Johan Rung and Alvis Brazma, "Reuse of Public Genome-wide Gene Expression Data," *Nature Review Genetics* 14 (2012): pp. 89–99.

30 Paul Muir, Shantao Li, Shaoke Lou, Daifeng Wang, Daniel J. Spakowicz, Leonidas Salichos, Jing Zhang, George M. Weinstock, Farren Isaacs, Joel Rozowsky, and Mark Gerstein, "The Real Cost of Sequencing: Scaling Computation to Keep Pace with Data Generation," *Genome Biology* 17, no. 1 (2016), doi: 10.1186/s13059-016-0917-0.

31 For the literature, see here and, subsequently, the references at the beginning of Part 4 of the conversation.

Rudy Raff and Denis Duboule, independently provided a theoretical solution to the problem. Raff, in particular, suggested that during much of early and late embryonic development the gene expression networks underlying the processes that give rise to organismal form are relatively isolated from each other in terms of function. However, during the phylotypic stage, when embryos from different species look similar, these gene networks may be highly integrated due to pleiotropy—that is, when one gene influences many different processes—which would constrain evolution's ability to change a gene's or network's activity in one context without also affecting another context (likely negatively). It was not really possible to test this hypothesis until high throughput -omics technologies became available in the early 2000s. Since then, several studies have used transcriptomics, where gene expression profiles are examined on a genome-wide scale, to demonstrate that embryonic gene expression profiles are significantly more similar during the phylotypic period than in earlier or later embryonic stages. These results confirmed von Baer's morphological observations at the molecular level and expanded the concept to invertebrates and plants, although they did not have enough statistical power to test Raff's hypothesis.

We discuss here an ongoing *in silico* study of the developmental hourglass, which tested the hypothesis that embryonic gene co-expression networks transition from more a modular topology before and after the phylotypic period to a highly interconnected one during this period. To do this kind of research, Hofmann and coworkers first searched public repositories for datasets containing the transcriptomes of vertebrate development across multiple embryonic stages. They then assessed the data quality and available metadata (e.g., experimental design, methods used) and communicated with the authors of some of the datasets directly to resolve inconsistencies and obtain missing information. This lengthy, almost year-long, process also included exploring various statistical procedures to standardize the datasets so as to make them comparable. In the end, developmental datasets for five vertebrate species, a fraction of the originally surveyed data, passed all quality control steps. The researchers then conducted a comparative analysis of the developmental trajectories of transcriptomic networks across species. To complement and extend this research, they used a novel *in silico* evolutionary model of a developing tissue to test different mechanistic hypotheses of phenotypic diversification. The results demonstrate how transcriptomic architecture is associated with the generation of phenotypic variation across species. With a poster of the developmental hourglass in view, we discuss these approaches, analytical steps, problems encountered, and the status of the knowledge gained. What does it mean to research *in silico*?

Acknowledgments

Thanks to Kathrin Klohs (Basel) for the editorial review of the four conversations, Birk Weiberg (Zurich) for coordinating the editing of the complete volume and for his contributions to the *Computer Signals* art project, and Susan Richter (Berlin) for translating the conversations and all accompanying texts in this volume. Valentina Vuksic (Zurich) shoulders responsibility for the artistic side of the project and participated in all of the group's discussions, aside from the one up on Rigi. We would like to thank her especially for her critical review and amendments to individual discussions. Michael Heitz took on this conversation between the disciplines as our publisher and was a great help in realizing the book; for this we thank him as well. To Christoph Schenker, director of the Institute for Contemporary Art Research (IFCAR) at the Zurich University of the Arts (ZHdK), we extend our thanks for the many years of supporting the research projects of Hannes Rickli, for his critical solicitude and generous logistic support in the framework of the Institute and the Fine Arts Program at the

ZHdK. The *Computer Signals* project and parts of this publication were financed by the SNSF (Project no. 169378). The editing of the conversation and the publication of this volume were made possible by the IFCAR (ZHdK) and the Research Committee at the University of Lucerne.

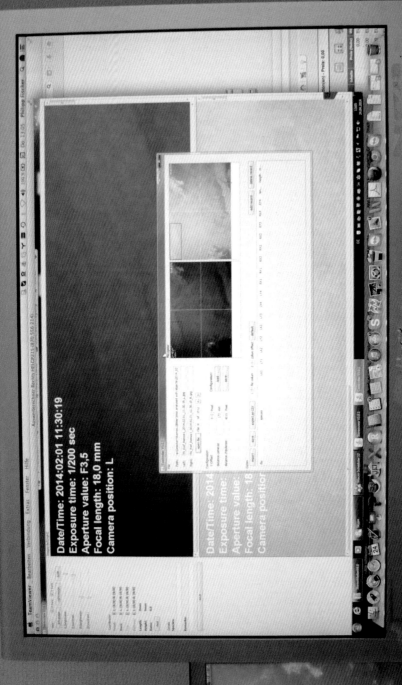

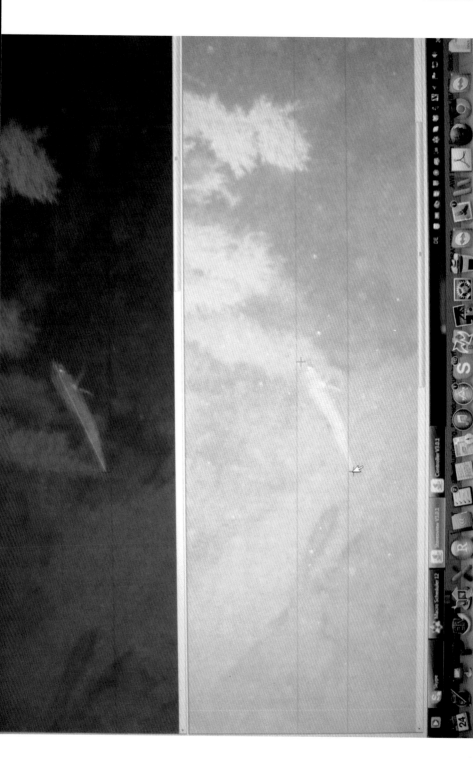

Data

Analog/Digital

Christoph Hoffmann Work with large amounts of data, data-driven research, the ever further-reaching data work—these are a few of the keywords we have been dealing with in recent years. One starting point for our discussions was the impression that we are experiencing a shift in research from the visual to the digital. Not a complete replacement—tables, graphs and composite diagrams continue to play an important role—but, increasingly, a data space is emancipating itself alongside these as a new experimental space. Philipp's fish acoustics project offers a clear illustration of how images are forced ever further into the background, first of all in data production.[1] In particular, I'm thinking of the transition from continuous video recordings to three-second snippets that was completed between 2011 and 2014.

Philipp Fischer I briefly want to explain why we did that. We urgently need the video analysis, that is, the behavioral information on the fish, to be able to analyze the sound data at all, that is, to match a certain sound to the behavior of an individual animal. There are two problems with this: First, we worked on the acoustic recording issue for quite a long time. But eventually we got this running. Ultimately, however, the overall number of the video recordings became so immense that we were no longer capable of even coming close to coping with the resulting data volume. Then we thought about how we could find a way out of this dilemma. We succeeded by optimizing the digital, computer-based analysis in such a

---

1   The subject investigated in this project is the inter- and intra-species communication between gray and red gurnards through actively generated sounds. Small groups of fish are held in a research tank for several days, and their behavior recorded continuously on audio and video. In the latest series of experiments, continuous recording has been replaced by event-based recording.

way that the fish acoustic signals were filtered during the online record-ing stream, and the target video snippets were cut out directly in real time. The development of this software, which allowed us to differentiate the acoustic signals coming directly from the experimental setup into background noises and fish noises, made it possible for us to reduce the video data such that the data stream became manageable.

Christoph Hoffmann You say that the visual data are indispensable for the evaluation of the acoustic data. Can you not imagine that the system might someday work entirely without images?

Philipp Fischer No, and that's because in this project the concrete object of our investigation is acoustic signaling during certain behavioral patterns. We want to find out what information content is coded in the acoustic in-formation, and we can do this only if we find out which action-reaction the fish are showing when they make a certain acoustic signal, a growl, a grunt, a knock. That's why we invested a great deal of time, more or less successfully, in learning how to match the acoustic signals to individual fish behavior. For a long time this was impossible. We received and re-corded acoustic signals from the test aquarium and found that they were fish noises. But when we had two or three fish in the video image at the same time, we did not know which of these fish had emitted a specific signal. Matching each signal to a certain fish, which meant acoustic tri-angulation in our case, was crucial in order to evaluate a fish's behavior in terms of the actions and reactions of the specimens in the experiments.

Hans-Jörg Rheinberger I don't believe in an arrow from the analog and vi-sual to the digital and numerical. What we are negotiating under the con-cept of the digital has to do with media, either recording or processing media. But everything that is digital eventually lands back in the realm of the analog, simply because we are analog beings and can only receive analog data through our senses. Ultimately, we always fall back on our senses. We are dealing with what we can receive through seeing and hearing. Our sensory organs may receive discretely, but first of all, dis-crete is not the same as digital; and second, our perception has always been synthetic.

Hans Hofmann   In the meantime, however, a transformation and a reduction in complexity have occurred. Of course, we can take in only visual or auditory input—or whichever input our sensory systems allow. But what we have then no longer has much to do with what was originally recorded.

Hans-Jörg Rheinberger   But the paper that you are publishing in *Nature* does work with images. And it works with narration; it uses all of the classical, traditional means. One always ends up back on this surface.

Hans Hofmann   That's true, only you can no longer directly recognize those data that were originally collected.

Philipp Fischer   To a certain extent, you're right. At the level we are talking about now, digitization is, in fact, only a medium. It simply means that I no longer have to take a ruler to measure an object on the image to extract the relevant information, for instance, to measure a distance of four centimeters between a certain fish on the image and another, and then to detect how this fish approaches the other when a signal is given, reducing the distance to three centimeters, and so on. Now I can extract this information from my digital image with the mouse because I am interested in the distance between the two animals. Ultimately, we extract numerical information from the images. For us digitization is a simplification, another tool that makes it easier for us to access numerical data.

Hannes Rickli   What's interesting is that the visual data are so comprehensive. They have, as you say, brought the system to the brink, so that filtering has become mandatory. As a consequence, fish communication is selectively recorded: only under a positive aspect, namely whenever a fish sound is detected. Another question worth investigating would be whether there are also similar situations in which no communication takes place.

Philipp Fischer   I absolutely agree with you. That is still our weak spot, which we cannot resolve at the moment, because our scientific approach is actually based on the acoustic approach, such that any information we extract from the video images must, unfortunately, follow the primary evaluation of the acoustic data. However, it is good practice of a behavioral biologist to prove that the behavior we observe and associate with acoustic patterns is not predominant even without the acoustic signals. If that were the case and we have 50 percent or 60 percent of a specific behavior even without the acoustic signals, then we may be wrong to associate this behavior with specific acoustic patterns.

Hannes Rickli So how do you measure that?

Philipp Fischer At the end, the entire video material has to be analyzed. But at the moment we do not have the slightest clue, nor any tool for going through the entire video, independent of the acoustics, to search for certain behavioral sequences of the animals.

Christoph Hoffmann Hans once said that the behavioral data were the most complicated data in his work. I assume that complicated means that these data are the "dirtiest," because here researchers have to make so many decisions on their own?

Hans Hofmann We have to consider several aspects. These data are the most difficult to collect, because they are multi-layered—the animals are moving in three-dimensional space over time. The data are thus multidimensional, whereas sequence data are one-dimensional. Moreover, there is not just one fish, there are many fish. This aspect and the work it takes to extract from a video image not only the XY-position of the many fish in space, but also their orientation, makes for very great cost and effort in comparison to the sequencing data we receive from the sequencing facility. However, this does not necessarily mean that the behavioral data have more noise.

Christoph Hoffmann With "dirty" I meant only that these data are based on human decisions.

Hans Hofmann Yes, that's true, if the behavioral data are collected on the basis of a traditional ethogram, an inventory of behaviors. This entails searching for various types of defined behavioral displays, defined somehow on the basis of our cognition. But nowadays we also have systems available in which machine-learning approaches are applied in order to define behaviors without any previous human input. This is quite interesting, as for the most part, the same behaviors are identified. Apparently there is no major difference between what we see and can agree upon objectively, and what is produced by machine-learning approaches. I can remember two papers: one on *Drosophila* flies and one on *C(aenorhabditis) elegans*—relatively simple organisms, although their flight behavior alone is quite complex.[2] Machine-learning approaches allowed the authors to identify a couple of additional behaviors that were not necessarily apparent to us. What was also interesting about these papers was that they presented hierarchies that defined which behaviors were more similar to each other than others. But I think that we are actually faring quite well with what we can recognize as a certain behavior and what I can show my students, that is, what we can agree upon. I have relatively few reservations regarding the building blocks of behavior.

Christoph Hoffmann I came back to this comment because it really impressed me. For someone who is not a biologist, it initially seems counterintuitive that behavioral data are the most difficult to collect. I would think: Well, counting behaviors is not so complicated. What's really complicated are the sequencing and interpretation of these data.

Hans-Jörg Rheinberger This indicates that we are dealing with an extremely diverse data concept. Of course, there are numerical data. But behavioral patterns can also be conceived of as data. Or, in ecology, even entire organisms can count as data. Talking about digital data conceals this variety and tends to reduce the data concept to what is digital. Data need to be defined in a totally different way. I believe we need a completely different framework.

---

2   On *Drosophila*, see Mayank Kabra, Alice A. Robie, Marta Rivera-Alba, Steven Branson, and Kristin Branson, "JAABA: Interactive Machine Learning for Automatic Annotation of Animal Behavior," *Nature Methods* 10 (2013): pp. 64–67; on *C. elegans*, see Eyal Itskovits, Amir Levine, Ehud Cohen, and Alon Zaslaver, "A Multi-Animal Tracker for Studying Complex Behaviors," *BMC Biology* 15 (2017), article 29.

Christoph Hoffmann   That's right. Only, one aspect of using these so-called complex behavioral data includes that, at this juncture—as Philipp's work shows as well—the work is moving toward automation. The attempt is to switch off "the human interface" from the outset and to perform direct pattern recognition instead.

Hans Hofmann   To become more efficient. High throughput and so forth. Yet, in my opinion, behavioral data are also the most difficult because they concern the very experimental design: What is actually the scientific question to be pursued? Many of the things we do in my lab start from the level of behavior. Then you probe downward or upward, or in whatever direction. I find the task of posing the right questions about the animal considerably more difficult than evaluating any kind of sequencing data.

Gabriele Gramelsberger   I am not familiar with your experimental system, Hans. Do you, like Philipp, digitally record the behavior of fish on video?

Hans Hofmann   In some instances, depending on the experiment, we film from above, and then the XY-coordinates for each individual animal are digitized directly; today we no longer have to tag the animals. We record the direction in which the fish look and how they interact. Then various algorithms can be developed, which can be used to determine whether the fish are actually interacting with each other or not. But when more complex social behavior is concerned, we still have undergraduates score it manually. That is what digitization actually means here. The videos themselves are digital, of course. Nothing is recorded on tape anymore. But, depending on the experiment, the behavioral data—we differentiate around 18 or 19 different behavioral displays for these fish—are still extracted in part by humans. I would love to automate this process more, but it is pretty efficient as it is. The data are coded on a computer keyboard. People with a lot of experience can do this nearly in real time. You can set the recording at 80 percent, 90 percent of the actual speed.

Christoph Hoffmann   I know from Alex Jordan, who is now working in Konstanz, that his research on social cues initially yielded quite surprising results. It turned out that the dominant fish were not the informants, but the subdominant ones, and that these, accordingly, accelerated the group's pace of learning. He then examined the data again, going backward all the way to the beginning, to the counting process. The data were scored again and then reviewed to see whether there might be a fundamental error in the whole history. For data generated "by hand" it seems to me that this is quite easy to accomplish. But what if these activities are automated? Can the experimenter still return back to the starting point?

Hans Hofmann Yes, in principle. The most important decision here is: Which of the intermediate steps do you keep when the data are transformed, or compressed, or extracted, or whatever else happens to them? What is stored? In some cases, if all of the intermediate steps were discarded, you cannot go back to the start. Or you have to start over with the original videos. However, the videos take up a lot of storage space. That's why some may say that there is no need to keep the videos. My approach is different. Storage space is so cheap these days that I always keep the videos as original data. Much of the data on the intermediate steps that take place during evaluation to produce the data file, for instance, the XY-coordinates, is then discarded.

Hans-Jörg Rheinberger So you keep in mind that it may be necessary to go back to the beginning?

Data Reproducibility

Hans Hofmann Yes, and we do that relatively often. We recently published a paper for which we evaluated eight different experiments performed in my laboratory since 2003.[3] The article is thus a kind of meta-analysis, although I had not actually intended to publish these results in this form.

---

3   Peter D. Dijkstra, Sean M. Maguire, Rayna M. Harris, Agosto A. Rodriguez, Ross S. DeAngelis, Stephanie A. Flores, and Hans A. Hofmann, "The Melanocortin System Regulates Body Pigmentation and Social Behaviour in a Colour Polymorphic Cichlid Fish," *Proceedings of the Royal Society B: Biological Sciences* 284, no. 1851 (2016), doi: 10.1098/rspb.2016.2838.

In these experiments we had always recorded a certain datum for a fish, which had been practically irrelevant until this time. At some point the question came up as to whether this aspect—it concerned the body color of the fish—could be explained in another context. I then suggested that we look at all of these different experiments to see whether we could identify any kinds of trends or patterns. For each individual experiment the data are pretty noisy, without any (statistical) significance. But when you take a look at this over all eight experiments, it is quite clear what happens. This is another way to go back to the data. For me this is very important. I know that it would not be possible in many laboratories, because the recordkeeping is simply not good enough to allow this kind of data archeology, even in one's own laboratory. This is a tremendous problem.

Christoph Hoffmann It seems there is a trade-off here between digitization and data handling. Philipp decided to replace continuous video recording with the storage of three-second snippets from the running stream because the amount of data had become too large and the material could no longer be evaluated. Hans, for his part, mentioned at our *Fragile Daten* conference in Berlin that the storage of sequencing datasets is a problem because they require a great deal of space. For him, the reaction would be to no longer save the data, but only the information about how the data can be acquired again—that is, the metadata. Here one sees how, as an effect of digitization, data gaps are produced that are quasi inherent to the system. Afterward Philipp is no longer able to access all of the images in a stream. The stream flows, it is gone. He must run this risk because the digitization of the experimental system forces him to do so. Would you agree that—put in slightly exaggerated terms—such experiments are intentionally moving toward information loss, because otherwise, the expense, one could also say: the concern for these data, is too high?

Philipp Fischer I find this point quite valuable. We had precisely this discussion in the *In situ ecology and technology* working group at the AWI (Alfred Wegener Institute). I would like to turn it around a bit, however. We decided, with a heavy heart, to record only these short sequences. We were fully aware that in doing so, we would be cutting out parts of the information that we might be able to utilize later. At our Spitsbergen monitoring station, we decided against it for our FerryBox data and all of the image data.[4] While we do not keep all of the data produced, we do keep all of the raw data, so that all data can be recalculated again with our algorithms. That adds a little more to the mix, as we can sometimes consider whether the raw data are the binary data that come from the sensors, or whether they are the actual values (for instance, the salt content). In any case we now have to be meticulous about saving all of the metadata. These are imperative in order to be able to get from the original data recorded by a sensor (i.e., the binary data) back to the ASCII data that we can read. If we do not have these metadata, that is, if we lose the encryption code, then the data can no longer be decoded into readable data. Safeguarding the correct codes to decipher data from each sensor at every point in time is at least as demanding as ponderously writing down all of the processing steps. Just recently we wanted to re-evaluate a dataset from the year 2012, but we were no longer sure which calibration dataset belonged to the sensor on that day of that month. Actually, now we are stumbling into another problem, which—in contrast to a shortage of storage space—is not directly and exclusively a consequence of digitization. We have to establish extremely consistent mechanisms to ensure that we are able to put the data back together correctly. In the meantime we have moved on to saving not only the data, but also the evaluation programs in MATLAB and in R with which we processed these data. At the moment we are still discussing whether to include the associated evaluation algorithms and metadata when we publish data in the major repositories like *Pangaea*. For us this is an important matter. To answer the initial question: We will not settle for more gaps in our data in the future. On the contrary, we will attempt to guarantee that all data can be recreated even after five years.

Hans Hofmann We have the same problem. Reproducibility is when one

---

4   Since 2012, the Alfred Wegener Institute, Helmholtz Centre for Polar and Marine Research, together with a French partner, has been operating the AWIPEV-COSYNA Svalbard Underwater Observatory in Ny-Ålesund on Spitsbergen. It consists of various sensors, an underwater stereo camera unit, and a webcam. Via a data cable, sensor data and photos are forwarded to the AWI Computing and Data Center in Bremerhaven.

takes data from the same datasets, performs the same analyses, and actually arrives at the same results. This is anything but trivial. The reproducibility of analyses is currently a huge topic for us in genomics. The idea is to save the programs that were used along with the data, or to establish permanent links to these programs, and then assign the programs a version number. The whole concept of version control is becoming crucial. If I open a paper on any kind of genomic analysis at the moment and take the data, download them from any kind of public repository, take the algorithms from there, the probability is very high (even if the analysis is described quite well) that my results will be different from those in the paper. This is a problem.

Philipp Fischer   Absolutely. We are working on a data portal that collects the data from Spitsbergen, but also from Heligoland, and immediately issues an error message if we are missing any metadata or associated algorithms that have not been stored. At our institute we are also currently discussing preventing our own data from being fed into our online repositories at all unless all of the information we need to process or handle these data later are stored in the repository as well. This would even mean that we would be cut off from our very own data if they were not documented comprehensively. According to our experience, this is the only way to build up sufficient pressure to make us active enough as data-providing scientists. Without comprehensive documentation, a screen message or a mail would be output with the indication: Collection of your temperature data from Spitsbergen will cease immediately; the repository is lacking the relevant algorithm.

Gabriele Gramelsberger   What you two are describing is a fourfold data concept. There are raw data, metadata, and data about the analysis method, and then the data about all of these data. The layers just keep piling up. When you ultimately speak of *the* datum you have to unravel all of this, because otherwise you have no chance of understanding this datum.

Christoph Hoffmann   In fact, it is striking how researchers' concern is shifting ever more toward the metadata. Doesn't this, again, pose the problem of the gap and what is lacking on a second level?

Hans Hofmann Metadata are a major problem. First the community must reach a consensus about which metadata are relevant. Even this is turning out to be not so simple. But even if a consensus is reached, it could be that in one, two or five years, someone would like to have all of this other information that was deemed not important when the compromise was made. This then raises the question as to what people get for actually making their data publicly available. In English this is called the incentive structure. Why should they spend hours or days putting together all of these metadata and submitting them to a public repository where they are only going to lie around, unless they receive something in return? As far as reproducibility is concerned, there must be agreement as to what is good enough. This is not about 100-percent reproducibility, but about when the results are considered to be robust. Different researchers with somewhat different approaches can analyze the data. The results will not be precisely identical, but they should be concordant.

Data Scientist

Philipp Fischer Reproducibility is a multi-layered matter. The reproducibility of our data is conditioned, on the one hand, by our experimental approach. When I decide on an experimental approach, I will always produce gaps, because—as Hans correctly said—I consciously decide to collect certain data and not to collect others. This is the normal scientific process, which requires scientific expertise. In this sense I will never get around gaps. On the other hand, a datum or a number I produce is my anchor. Described in that datum or number is my level of knowledge, with the idea of the experiment coded in the first numeric value. Starting with this numerical value, we aspire to the complete reproducibility of the data. These are two different things. We reduce the idea to the nucleus number. To get from this idea, the scientific question, to this number, great scientific expertise is needed. But starting from this number, reproduction should succeed at a rate of 100 percent. Only then do I have a correct data structure. In order to ensure this, we are currently intensively discussing a new curriculum within our university program.

We need a new type of scientist that we may call a "data scientist." The US is leading the way; we're not so far in Germany yet. With the term data scientist we mean that this person concerns themselves exclusively with data in science. On the one hand we then have the scientific process of research, in the course of which concepts are put forward in order to acquire data; on the other hand, we have the data scientist who retrieves, produces and reproduces these data and erects multilayered data models. We want to encourage this: the data scientist and the natural scientist. I am not trained to deal with large amounts of data. I pick up these skills only with a great deal of time and additional effort. But in principle, I am not trained for this.

Hans Hofmann   At our university we have a Department for Statistics and Data Science. They were recently awarded a training grant in data science. One of my tasks is to establish a new graduate course: BIO 382K, Introduction to Biology for Data Science. By now this topic has arrived in graduate training, and is gradually entering undergraduate training as well. A few years ago, we made the decision that nobody can finish with a Ph.D. in most of the life science graduate programs without a strong foundation in data processing and data analysis, and thus big data skills and the like. Everyone who studies with us has to have these, even the molecular biologists, biochemists and ecologists. I think we are about 85 to 90 percent of our way to this goal. A few manage to slip through every year, but not very many.

Hans-Jörg Rheinberger   On the one hand, the people who work in this area have to have sufficient connectivity. On the other hand, however, differentiation is clearly taking place. In the historical perspective, at some point, people working in the natural sciences had to have recourse to technicians, because the technologies became so enormous that the scientists themselves were no longer able to fully master them. Now the data space is expanding, and for that specialists are needed as well. This is trivial in principle. I believe that the decisive point is that the data space is acquiring an ever stronger presence, ultimately becoming an experimental space, one that is incredibly multifaceted and needs new skills to be mastered.

Hans Hofmann One should be careful with prognoses, but it would be desirable for the biologists, the life scientists, ultimately to be data scientists themselves, so that they are not dependent on others. In my opinion it is still important that those who have learned to pose questions and design experiments—be they *in silico* experiments or experiments that actually take place in the laboratory—, that the same researchers who generate the data are also able to analyze the datasets. At any rate, this is the decision we made. On the other hand, the scientists from engineering, from physics, from computer science, are coming in now; everyone wants to perform research in biology. We want them too, for they have approaches and expertise at their disposal that we are lacking. These scientists move very differently in this data space. We have to teach them biology, and that is my task with this new course. So it's moving in both directions. Seen from the historical perspective, this is nothing new, of course. Every now and again there is a wave of physicists or engineers who come to biology, with varying degrees of success. One such movement was cybernetics, but there are further examples of this. It's quite similar now. In physics I know several grad students and postdocs for whom it is simply no longer interesting to be a physicist in the sense of working at some accelerator and having their name on a paper along with 10,000 other researchers. They consciously decide to go into biology, where their name is on the paper with only ten or fifty others.

Philipp Fischer I would agree. In principle, the approach is correct: our biologists have to be capable of handling data. In our disciplines, too, we have many persons who can do that because they have already been working with these large datasets for many years—above all in atmospheric research or in terrestrial ecology, and somewhat less in ship-based aquatic ecology. This is not the case in our special field of behavioral research, nor in coastal research, which has become relevant in recent years as it relates to climate impact research. It really depends on the discipline. For my students—that means students from everywhere, be it the US, Europe, or Africa—I unfortunately cannot confirm that the majority are able to deal with data. I am actually pleased when students are able to work with Excel, and there are perhaps ten to fifteen percent who also know how to use MATLAB or R. Nevertheless I believe that we need specialization in these areas. I would not like—on this we are in complete agreement—for us to have scientists sitting in a computing center who do not have a clue about biology, and who then take our data and evaluate them further. But in the future I hope that not only the scientists who research in the laboratory get credit, but also those who make the effort of working more intensively with data. This is our main problem. The scientists who sit at their computers and work mainly with data are always quite poorly evaluated in terms of academic credit. At our universities and research institutes we at least attempt to convince scientists that working with data is also attractive.

Hans Hofmann I think genome research has made a bit more progress, although this discussion is taking place in our field, too, of course. There is a conflict between those who produce data and those who download data from one public database or another and then evaluate them. The latter are generally referred to as "data parasites." I believe I read in the *New England Journal of Medicine* that a few scientists who want to keep their data have banded together. They said that they do not want to share their data with others who then simply publish a slew of papers from their data. There has been massive pushback on this. But at my department—we've been doing this for ten, fifteen years—we hired a number of scientists, faculty, who pursue computational biology and bioinformatics. They have no wet labs; they only collaborate with other researchers or download large datasets. In my laboratory I also have scientists who do only bioinformatics. We have a large-scale project in which we do not produce any data of our own, but receive all of our data from databases. And I see that the scientists who do these things and gain new insights actually do receive precisely this credit for this work. But it is a slow process, and I believe some institutions are a little further ahead than others.

Trust

Christoph Hoffmann At this point I would like to address a further aspect. It has been mentioned already how complicated these data are and how much knowledge about a dataset is needed in order to work with it. I ask myself whether the strong link binding data to the location of their generation can be broken. Does it make sense to set up large repositories with the intention of stimulating scientists to share their data? What has your experience been? Do you work with data from other researchers whom you don't know? This interests me because I see a jumble of uncertainties concealed behind the demand for data sharing advanced by science policy. We already addressed one possible way of breaking down this jumble, when we talked about the question of how comprehensive metadata must be in order to be able to work with a dataset. Another possibility is designated by the word trust: a word many perhaps shy away from.

Hans Hofmann Let's start with trust. I have no problem with the word. Trust is the foundation of science. If I cannot trust the persons who perform a certain experiment, there is no basis. I have to be able to trust that the experiment and the results they report are correct to a certain degree. This does not mean that the interpretation is correct and that ultimately everything will be right. Scientists develop a reputation as a consequence of this. Some of my colleagues have a reputation that leads me to treat the data they produce with perhaps more skepticism than those from other scientists who have a better reputation. This is quite well known within any given community. Now, concretely to answer your question: In my laboratory we use many data that we did not produce ourselves. First, at some point everyone in my laboratory, be they grad students or postdocs, or sometimes even the undergrads, makes analyses of data that were produced by someone else in the laboratory. Even in these cases there are sometimes misunderstandings and problems: Something is not right, or not completely documented. Then it often turns out that I and my institutional memory are needed in order to fill such gaps. Second, of course we use data from our collaborators, from researchers whom we know well, whom we trust, with whom we have a certain relationship. And third, we use data that are simply available in these public repositories, deposited there by researchers whom we do not know personally. In this case rather nebulous concepts apply, such as reputation. Or we contact colleagues, because—as happens regularly—we need additional information. It then turns out that some are very generous about what they share with you. They answer quite promptly, while others do not. You don't necessarily know why, but at least subconsciously you do the math. Generally our experiences with this have been very positive. It is actually quite unusual for us to say: Now that looks funny. Or: We have to be skeptical here. However, it is true that we perform our own checks, of course, even if they take place within the laboratory. For instance, someone needs hormone data. Blood samples were taken from animal subjects, and three, four different steroid hormones are to be measured. However, the scientist interested in hormone data does not perform these measurements herself, but another, who is doing an assay (and thus a whole slew of such measurements) does so and adds these 30 or 40 samples provided by the colleague in the lab. The one person must then trust the other that the data are actually reasonable. I expect of my staff, however, that they look at all checks and references themselves to ensure that these data truly make sense. This is the same thing we do when we take data from public repositories. We first put these data to the test and consider whether they might, under certain circumstances, include a systematic error. Of course, this is not always entirely possible, but you get a good picture.

Christoph Hoffmann Don't you have the feeling that you're building on quicksand with these data?

Hans Hofmann No, if there are data of that kind, we keep our distance. We have the problem that the technical developments are progressing so rapidly that we often have a compatibility problem. In a relatively large project that is only computational, in which we did not produce any data ourselves, we are dealing with data that were generated with what are known as microarrays, and with data that were generated with what is called next-generation sequencing. Something along the lines of a Rosetta Stone is needed to make these compatible with each other. This is anything but trivial. If you cannot manage this, you have to throw away a major portion of the data that are, in principle, at your disposal, which you don't want to do. Or you have to analyze them separately. That is the real challenge: the different technical platforms that develop further so quickly.

Hans-Jörg Rheinberger Does that not depend a bit on the area and the community? I'm thinking of the work by Sabina Leonelli, who spent several years with the groups that have been collaborating on a shared model organism, *Arabidopsis thaliana*.[5] These communities of perhaps 500 or 600 people spread all over the world used shared databases. There is a repository into which more or less everything about the model organism is fed, but at the same time the repository itself is constantly being worked on. This raises two fundamental questions: Which criteria must all of these data fulfill in order to be entered into the pool at all? And how specifically must they be indexed so that the individual working groups, each of which is addressing its own issues, can access this pool expediently? This is a question of generalization and specification, of global scope and local exploitation at the same time, and the trade-off between these two poles is, I believe, the decisive point.

Hans Hofmann And the question is not resolved; this is—as they say—a constant work in progress.

Christoph Hoffmann At the same time, it is apparent that the quasi-technical solutions of the circulation problem, from barriers to including datasets in repositories, to the rules about the metadata to be supplied along with them, are never sufficient. Hans just affirmed this once again. One has to contact the scientists about their data, perhaps because something is missing, and even in one's own laboratory, the staff has to make inquiries about the data they are using. It doesn't work without communication. That's why I asked about trust.

---

5   Sabina Leonelli, *Data-Centric Biology: A Philosophical Study* (Chicago: University of Chicago Press, 2016): pt. 1.

Philipp Fischer I agree completely with Hans, that one has to have a certain degree of trust in external data. At the same time, however, a healthy distrust is needed as well. The mere fact that there is no logistics enabling us to produce all data ourselves means that we necessarily have to rely on external data. If the subject is climate impact research, for instance, we need more or less comprehensive data from the Arctic and/or the Antarctic: temperature, salt content and whatever else. For this we have to rely on the repositories of the USA and other countries. Whether or not we trust these data is not a question that is posed at all. We have no other alternatives, because these are the only data available at all in order to calculate the values in certain models. We use quality management algorithms, of course. Last week I even sat down with a colleague who actually plots the data he takes from external repositories by hand to see whether they make sense. Inversely, a healthy degree of trust is also part of this. Now we are once again taking recordings from the *Polarstern*, starting from Bremerhaven—North Atlantic, South Atlantic, then toward Cape Town; once these are published we will work with them. In this case, the data provider has to enjoy a certain reputation, however. For the *Pangaea* repositories, where our data are stored, we have a very strict rule. We submit the data to those persons who are responsible for the repositories, they check these data independently for consistency, and only if they pass this check are they published. Another example: We have very important, unique data on ocean acidification off the coast of Spitsbergen. Now a colleague from France says: In order to use these data for further calculations, we require the silicate data from the target area for calibration of the sensors. So we asked around: Who, by chance, may have collected such data? From the Netherlands we then received the message that they determined such values in their weekly manual samples over the last year and a half. This is a typical co-utilization of datasets that are rare or very costly to collect. We know the colleagues, they measure very accurately, and some of their data are also analyzed by us at the AWI. One example from my own group: In connection with the data from Spitsbergen we also measure the age composition of the fish populations. For this we dissect the otoliths, the tiny ear bones, and count the growth rings, just like for trees. As a result we can say, for example: This animal is six years old. Sometimes the rings are not easy to detect. But if we classify the fish incorrectly, we have a huge problem. If we say, for example, that the animal is six years old, but in actuality it is only two years old, we postulate that the animal has been sexually mature and producing progeny for three years, which could potentially be important for the size of the fish stock and also for the fishing quota as well. This is why we have the analyses performed by three persons: I have one person analyze the otoliths. Only I see the age data for each individual fish that are measured by this first analysis. Then a second, independent person from our group is tasked with analyzing the same dataset. Then I decide whether or not their results match. If they do not fit together, I ask a third person, who, again, has no idea of the first and second result and this person analyses the otoliths once more. Finally I then decide which age is correct, or whether the data collected must be discarded. That's what I mean when I talk about the balance between trust on the one hand and healthy mistrust on the other. We need these procedures when the data are highly critical.

Hans Hofmann   Here we take our inspiration from Ronald Reagan, who is supposed to have said: "Trust, but verify!"

Accuracy

Gabriele Gramelsberger   That fits well with the question I have. I think that over the course of time, a data-critical awareness is developing in the research institutes. This struck me during my research on climate models and data. Satellite data are scrutinized especially critically. The reason for this is that they are loaded with theoretical presumptions (theory-laden) and that the algorithms involved are so complex that one cannot always trust such data from other institutions. This kind of theory-laden data is checked in house first, to determine whether they can actually be utilized. There are researchers whose only job is to perform quality checks on satellite data from other sources. This data awareness is exciting from a philosophy of science perspective.

Hans Hofmann What do you mean by theory-laden data?

Gabriele Gramelsberger These are data which researchers initially do not trust at all, because the data gathering methods are so complex that one first has to understand what assumptions have been put into them. And then there are data about which the researchers know that they are still "empirical" enough that they can be trusted.

Philipp Fischer It is a two-layered problem. For satellite data I know precisely that with the algorithms only a certain accuracy can be achieved at all. If I know the scattering around the actual value—that is, the accuracy—then I can deal with the data. If one takes the temperature data, for instance, it makes a big difference who needs them. If an oceanographer needs these data, she would like to have an accuracy of, let's say, at least 0.02 degrees Celsius. Otherwise she cannot utilize the data. If I take these same temperature data for my scientific questions, I demand an accuracy of 0.1 degrees or perhaps even just 0.5 degrees Celsius. Otherwise these data are completely irrelevant for me. The number of algorithms applied before I can use the data is one source of uncertainty about data quality. Yet it also depends who uses the final data product. At the moment we are experiencing ever more discussion about how to even calculate accuracy in a meaningful way, which is definitely not trivial. Unfortunately, in our data we still have classification through data flagging: For this the data are subjected to classification, for instance from one to five. One is for the raw data, two for the probably good data, three for the good data, four for the probably bad data and five for the bad data. These five classifications are widely published and thus recognized throughout Europe, and many of our colleagues work with these quality indicators. For us they are complete nonsense. Just imagine for a moment, I am calculating statistics, and in my statistics I don't say: Aha, I have an uncertainty factor, a p-value of 0.5, but rather: The p-value is probably good or probably bad. This is why we are currently working to systematize the real scattering range of data. When I have a value I want to know: In what range of spread is the real value with 95-percent probability? I would like to have the information about accuracy, and I would like to have information about precision: How does the value move around the actual value? Only such precisely characterized data bring real progress.

Hans-Jörg Rheinberger What also seems important to me—and this may go back to my own experience as a scientist—is that, in the first instance, the step of data collection is decisive. The data have to come from somewhere. During *in vitro* experimentation I had the experience that the possibilities for processing at one's disposal are usually a magnitude more precise. If you have to pipette microliters or half-microliters in such an experiment, when you've done that ten times, you have an error margin in the whole process. Afterward you insert the sample in the scintillator, in the machine that counts radioactivity, and this scintillator can measure every sample for you to any place after the decimal. But all of this precision is a waste of time somehow, because the upstream imprecision is much greater. For me, the awareness for what actually happens at the point of collection seems to be a prerequisite for everything else.

Hans Hofmann At least our training teaches us fairly well that measurement methods have different levels of precision. Just because I achieve high precision with one measurement method this does not mean that I can actually exploit this precision. But there are always negative examples of people misinterpreting this. For me this means only that we have not done our job when we trained those people.

Data Visualization

Christoph Hoffmann I remember how Hannes once said that he, too, can no longer cope with the flood of data—the data he himself collected from the research infrastructure in Austin, and the data supplied to him by *RemOs*. Initially this was a storage problem. But I think it is also a persisting aesthetic problem. Is this overload, put bluntly, due to the fact that there is no appropriate aesthetic form for the data? Are you entering unchartered territory with this?

Hannes Rickli Various factors play a role here. The first question was indeed about what kind of data handling would be suitable in order to arrive at formats that can be output for further processing. For Austin, for instance, it turned out that the conversion of the audio and video data into "films" that could be played back came at a huge computational cost. At the same time, we had institutional difficulties with the ZHdK, because access rights had not been sorted out, which was leading to server interruptions. What is more, the IT basics were lacking. Just like all of you, I notice that there simply aren't any standardized solutions for this. First off, the ZHdK had to learn to operate a server to which so many data are uploaded every day. As far as the audio data from Spitsbergen are concerned: Here we have just reached the point at which we are trying to get an overview of the various signal recordings. First we have to decide what scale we actually want: Do we want to proceed on the basis of events, or on the basis of time, that is, give an overview of three, four years of data? For audio this is not simple. While the signals can be visualized, one has to be able to go into detail quickly. When the signals are rendered in the overview, specific details are simply not visible. This is precisely the point of the audio panorama that we made for Hans in Austin. The recordings cover eight stations over 24 hours. At the moment every individual file can be viewed, but then the other situations are missing as benchmarks for each file. So, the question of simultaneity. Now I found a nice example for how this can be solved purely visually. For the Olympic Games there was this program called RealPlayer: nine channels with all simultaneous events, which were marked with an icon. You have a cursor, you know where you are on the time axis, and can jump into any channel as you wish. Perhaps this would have been a possibility to visualize simultaneity. For these acoustic things, sometimes the data-specific aspects are not relevant, not every individual signal that can be heard, but rather something more like intensity.

Christoph Hoffmann One thing I find interesting: You respond to the question about the aesthetic with technical answers. You speak of the difficulties in managing the flow of data or establishing simultaneity. But you did not give any reasons, for example, for why you inserted the recordings from Austin in an audio panorama. Is there perhaps a kind of logic that is dictated by the existing data, and from which the form of representation results more or less mechanically?

Hannes Rickli No, you're looking at it the wrong way. It's a situation that is new for me. I haven't actually done such work before. Therefore it is like a test, to establish a coherence between the data or the object and its form. But this is preceded by a concept: The point is to rematerialize aspects of this data production. I find all of these abstraction processes—from fish to dataset in the computer—extremely interesting, and I want to keep precisely this chain reversible. For me that is an interesting aspect of biology. One proceeds from a concrete organism, which then dematerializes into scientific work. My question as an artist is: How can I make this material basis that lies behind these many steps, these transformations—the infrastructures, energies, matter, etc.—how can I make it tangible? For the exhibitions *Videogramme* (Helmhaus, Zurich, 2011) and *Fischen lauschen* (Schering Stiftung, Berlin, 2013) I had videos and sounds as raw material. My artistic concept was to reconstruct the situation that yielded the images and sounds. I built the installations such that the devices that were involved simply took over the inverse function, so where the camera had been, I placed the projector, etc. That's not so simple in the Texas panorama. Therefore I am searching for possibilities to rematerialize the relation between sounds and images so that it can be experienced spatially.

Christoph Hoffmann Would you then say, Philipp, Hans, Hannes, that you are facing the same problem: Transformation of digital data into a sensory, sensually palpable form, albeit with different objectives? Does the step from dataset into representation present a problem for the sciences?

Hans Hofmann Visualization is definitely a major challenge. How can complex, often multivariate, multidimensional relationships be depicted visually in a way that makes them tangible for our brain? By now visualization has become a special field in data science. We offer courses in data visualization as part of our training program. But I am considering the extent to which there is actually a difference. We generally have concrete questions for the data. Either these are fueled by the hypotheses, or we conduct a more exploratory search for certain patterns and interesting relationships. For Hannes, by contrast, it's not clear to me how concrete his question is. It's a third layer: He does not seek patterns, and he does not test any hypotheses: he wants to make a process visible and tangible.

Hannes Rickli That's the problem, I'm interested in the process of data transformation, which can hardly be captured in a result.

Hans Hofmann For me, this process appears somewhere in the methods section. It is important, it must be reproducible. The relevant information, the metadata have to be there. But when I tell my story, write my paper, or give my lecture, the process itself is not relevant, it shifts to the background. Unless my research area is to develop methods, which is generally not the case for me.

Hans-Jörg Rheinberger That brings an anecdote to mind: When Bruno Latour was preparing *Iconoclash* at the ZKM in Karlsruhe, I actually wanted to contribute a short film to the exhibition. The intention was to visualize a scientific working procedure: What all is involved in coping with the technology of radioactive labeling? This was the question. My colleague from art history, Peter Geimer, was also part of the project. We sought out a scientist—truly a highly intelligent guy—at the Max Planck Institute for Molecular Genetics, and I knew that he had years of experience in working with these things. We wanted to realize the film with his help. We spent an entire day in discussions, but it was not possible to convey the problem to him. He always wanted the focus to be on the visualization of his results, his ribosome models. It was impossible to make him understand our question. It was like an experiment gone wrong. The film was never made.

Gabriele Gramelsberger Let's back up. As scientists, you are interested in the representationality of data. The data represent something that produces results. Hannes is interested in the materiality of data, that is a completely different perspective on data and their transformation.

Hans Hofmann   But it is not as if the materiality of data does not interest me at all. Their materiality is a challenge for me in scientific practice with regard to the documentation, storage and public availability of data. For me this is, however, entirely separated from the formulation of any scientific questions.

Hans-Jörg Rheinberger   What is also interesting, when I look back over the last two hours of our discussion, is that you are indicating precisely the critical point, but so far you have not given a single concrete example from the spectrum of problems you're working on.

Hans Hofmann   That's true. Earlier, when I was talking about the project for which we use only data from other laboratories, the hourglass project (see part 4 of the conversation), I even had at the back of my mind whether I should explain briefly what that project is about. But I told myself it would only take us off on a tangent.

Christoph Hoffmann   When I asked one of the collaborators of this project to explain the hourglass problem, she started opening windows on the screen. One, two, three, and at some point there were too many for her. She turned to the board and drew me the project—she didn't write it, she drew it. Hence the question about visualization. Visualization serves not only to pass on information; first of all, visualization facilitates self-understanding in the process of data analysis. One draws in order to find out what one wants to know. Although the intentions of Hans and Philipp ultimately are clearly different from those of Hannes, the process of self-understanding could be comparable, beginning when one steps out of the data space—Hans-Jörg also called it the experimental space—and searches for a visual depiction. How long does scientific work remain in the pure space of the computer and the software? When does one leave the data space and begin with visualization?

Hans Hofmann   That never happens separately; the two are always linked. The data analysis pipeline, all of those algorithms, all of this is visualized when I communicate it, and just as well when I think the matter through and consider what the next steps are. The very names of these software programs are often quite sensual. In transcriptome analysis, for instance, there is a standard toolkit, it's not quite so popular anymore, called Trinity. It goes practically from the caterpillar to the pupa to the butterfly. Those are the three parts, and they are even designated visually with the corresponding logos. If you then begin with these shotgun sequencing fragments and work through the three different main steps in sequence (each of which is divided into many individual steps), a transformation actually happens every time, it's like a metamorphosis. Just like when I ask: What is the flow diagram or the concept map of the entire analysis? The very expressions say it all, everything is visualized in some way. When I sit with someone in my office and we think about an experiment for the first time, we develop it on the whiteboard. Everything is done graphically, from the execution of the experiment to the analysis of the data, to the end, when a working model is drawn, which then may be included in the paper. With all of the question marks and the information we learned, it is a visual process through and through, which is completely entangled with the data generation and data analysis. I can't separate it out.

Hans-Jörg Rheinberger That means that the data handling also doesn't happen on the digital level. It is just as present visually.

Hans Hofmann Yes, in our heads.

Hans-Jörg Rheinberger As is the object.

Hans Hofmann What's more, for the majority of genomics, as far as mapping and the like are concerned, graph theory is important, which is *per se* highly visual.

Philipp Fischer I would like to pick up on a point Hans mentioned earlier: There was a time in my career, above all the time in Konstanz, during which I performed purely hypothesis-driven research. The acoustic experiments with gurnards stand for such hypothesis-driven research; in fact, all of my behavioral observations of fish and crabs were hypothesis-driven. Today my work is more explorative. I proceed this way because I work with my Spitsbergen data in a field about which almost nothing is known, so that we cannot even propose any hypotheses. That presents a few problems for us. The first article we published on this consists at its core of the visual description of our results.[6] But we were surprised ourselves, because we never would have been able to formulate a hypothesis which would have adequately described the results that we obtained from our actual research. The visual description of the data is extremely important for us, but it goes in a slightly different direction than for Hans. My data work takes place on the computer, and—presumably in contrast to the field of genomics—we have no ready-made software to evaluate our data. I write the programs myself, and at the end of the program is the graphic output of the results. Since our data have become so extensive, I am not confronted with the results *in toto* until the moment I have the graph in front of me. In fact, sometimes I do not recognize the result until I see the graph. This is why I regard data evaluation more as a flux. Let us look at the data space as a cube. I am sitting at one end of this cube and looking at its surface; an image is drawn there, and this image shows my XY-graph with a bar graph. Now anyone sitting at the other end of the cube has a different presentation and also a different materialization of the data. For me this is not so different from Hannes's work; also, because we have been discussing at the institute, especially intensively in the past month, how we can not only process our data visually such that we can understand them from the scientific perspective, but also how we can present our data from the polar stations to the public in a way that even the "common citizen" can understand. This discussion became quite confrontational. But in our department and at the institute, we decided that this is part of our scientific mission, which, by the way, brings us back to the credits. They are earned through a purely scientific examination of the data, and, increasingly, for data analysis as well, but we do everything else just for fun. When I prepare my data for the public, I am probably not so far away at all from Hannes's question formulation, question mark, whatever it may be. Hannes looks at these data—or so I imagine it—from a certain artistic perspective. The baker outside who

6   Philipp Fischer, Max Schwanitz, Reiner Loth, Uwe Posner, Markus Brand, and Friedhelm Schröder, "First Year of Practical Experiences of the New AWIPEV-COSYNA Cabled Underwater Observatory in Kongsfjorden, Spitzbergen," *Ocean Science* 13 (2017): pp. 259–272.

looks at our results views them from the perspective of a curious, or perhaps even interested, citizen.

The Research Object Form

Christoph Hoffmann What strikes me about Hannes is that he does not provide a new interpretation of your data, but rather, he lays down a second track. He uses your data, and the data he collects himself in the context of your research, in order to tell a parallel story.

Philipp Fischer Ultimately, that's what we do, too. Just look at the hole in the ozone layer, or, as far as I'm concerned, the North Atlantic Oscillation. In these cases, colleagues acquired data over decades with an entirely different objective. At that time the purpose was to research changes in the oceans: Temperature, salt content, simply to determine and monitor a dynamic trend over a period of 100 years, with the question in mind: What does this trend actually tell us? Then suddenly the suspicion arose that we have climate change, and all of those data of 100 years were re-analyzed; also a second track with a completely different research question, albeit within the same science.

Hans-Jörg Rheinberger There are clear parallels between the problem as it presents itself to Hannes and the scientific approach. It is the unprecedented. You cannot just anticipate the endpoint of your journey; you're caught within a search process. In terms of structure I would see quite clear parallels there. But the question is different. Hannes is not interested in contributing a small increment to oceanography or behavioral research. He is working on a different level. But he is confronted with same problems from an epistemological and also an aesthetic perspective.

Philipp Fischer I would like to react to this statement. I do not see Hannes in the stream from which we tap our data. Rather, I see Hannes with an experiment of his own, and we—that is: Hans and I, our work—are the experimental system *in toto*. We are the guinea pigs. That's what I've always said when I wanted to explain to someone what I actually do in the group. Hannes looks at my research and extracts from it the basis for his research.

Gabriele Gramelsberger Hannes is a "data parasite" with his own sensors in the experimental system. That is truly unusual.

Hans Hofmann But he also produces data of his own. To that extent, he is not a data parasite. I see myself less as a guinea pig. Rather, I see myself, my work—let's say—as a solar system or a continent or a river system, which Hannes observes from a distance with various telescopes and other measuring instruments. He is more of an ectoparasite, not an endoparasite.

Christoph Hoffmann Yet the step towards visual representation appears to be a major problem. This step is what occupies Hannes, it is the central aesthetic step. I agree with Hans-Jörg: In this point there is a parallel between the scientific and the artistic situation. Hans pointed out that by now there is a separate branch that deals with the visualization of data. His and Philipp's descriptions of data analysis show how many elements of surprise can be found in the visualization of data. However, it seems to me that in the sciences today it comes down to having a series of standardized practices at hand in order to deal with the data. For Hannes, the step toward depiction itself appears to be extremely critical.

Hannes Rickli There are always established solutions in art. At the moment it is extremely popular for the processual nature of a situation to be depicted via investigations and then perhaps with a couple of selected objects. But that is not my goal. You can depict your research as research better than I can. I am interested in how the research comes about—particularly at those critical points where technology interfaces with theoretical concepts and the like.

Hans Hofmann Which generally takes place in the background these days.

Hannes Rickli And it is not your job to address and discuss this. I would like to understand how these various aspects interact with each other. To do this, I first have to develop a form. The form itself is an object of research. What concerns me is—in Hans-Jörg's words—the shift from the epistemic object to the technical object, and how the technical object then becomes the epistemic one again. These technical objects would be the media, infrastructures, energy, the bioactivity that corrodes the sensors in Philipp's case, storms, icebergs, etc. I have been confronted with the technical aspects; I did not seek them out. Your system, Philipp, showed me how precarious that is. How closely interconnected the possibilities of the internet are, but also how dependent they are on a functioning submarine cable. Or on a functioning power source. Or a power outlet, a plug, plug connectors. For Hans that is probably much more standardized for the most part. There aren't these frictions, but probably others instead.

Gabriele Gramelsberger I find this point quite interesting. I see the difference between Hans and Philipp so, that Hans is in the laboratory while Philipp actually has his laboratory in nature, outside, and measures there. What Hannes just described is found in climate research as well. The supercomputers they work with are a laboratory within a laboratory. For instance, supercomputers need their own "climatology," in which they are enveloped, for the data flow to be generated at all. This is now becoming a subject of discussion itself. In the context of green IT there are projects in which the energy performance of the algorithms is measured, to find out which algorithms use more energy. What kind of a laboratory is emerging there? This is not a scientific lab, it is a data lab. A third space is opening up here, and this is the space that interests Hannes now.

Christoph Hoffmann I would like to stress once more: Hannes just described that there are standard forms with which simultaneity or seriality are depicted today. That can even be recognized in television series like *24*.

Hannes Rickli The split screen, for example.

Christoph Hoffmann These are things you have in the back of your mind, I presume. But now the point is to find your own form for your material? You know what you want to show, but you are seeking a form?

Hannes Rickli   No, that's the wrong way of looking at it. If my purpose was to depict Hans's or Philipp's research process, I would have artistic resources at my disposal. That is quite popular at the moment, and goes by the name of research art. I could arrange certain documents, extracted from the process, in order to produce an image. I would then say that the image is a representation of the process. But I am actually interested in the process in its materiality, in its temporality, for instance, and in its energeticness, that is, its intensity. That is the essential element. These qualities are not aesthetic, they are qualities of perception. They are affective qualities. These things interest me, and sometimes I really take them to extremes. When I put one of Hans's aquariums in the context of a thermal power station or a fracking platform, these are, of course, completely disproportional. The platform does not drill for the cichlid in the lab. But it engages me. I am on the lookout for ways to restore these relationships. In my view there are no prefabricated forms for this. The split screen would be one element that I probably will use, but perhaps there are entirely different forms. That is why I brought the issue of rematerialization so strongly to the fore in the new project.

Hans Hofmann   The whole thing sounds quite exploratory, like a newly discovered continent being sailed around for the first time. First you follow the coasts, and gradually you start making excursions into the interior. Then perhaps you fly across it at a high altitude with a low-resolution camera. Then you see the mountains and river valleys, and thus you slowly move forward, step by step. That's my impression.

Gabriele Gramelsberger   I find the term data panorama, which you often use in this context, very interesting. You generate another data panorama of the data space. The data space may be the same, but you generate an entirely different panorama from this data space than do Hans and Philipp. I think you know quite precisely what you want. The question is: What kind of data panoramas are we actually generating now as philosophers and science historians? What kind of a data panorama of your data space is it? This is the bridge between the various projects and views.

	Vorinkubation: 1. step					Hauptinkubation: 2. step						Translokation						PM-Re-aktion	Bemer-kungen
	Mix I	p(U) 3.75 mg/ml	tRNA type:	H₂O	TMNSH	ticos 8-30 pMol/aliquot	Mix II	[nur enzymat. System] GTP/PEP TU PK	TU	[Puf-fer]	H₂O	tRNA type:	GTP/PEP 1 Teil PK 2 Teile	Ali-quots auf-teilen	TMK	EF-G	PM in Bdg-Puffer	
per quot	5	5				103.7	20					Σ5	10	60μl	(5)	(5)	(5)	
per mal-satz	45	45	Σ45		Σ90	Σ90	180					Σ45	70	60 60 60 60 60 60	5 5 5 – –	– – – – 5 5	– 5 5 5 5	

Exp (316) : Rohdaten

Ph. R2 Exp 316

Software

Automation of the Observer

Gabriele Gramelsberger   The most exciting aspect of the topic of software, I find, is that the observer is replaced by software. I call this the automation of the observer: The activities that were originally conducted by researchers are delegated to software. How long have you had experience with this? When did you write your first software program, Philipp?

Philipp Fischer   I cannot say exactly. The software for the fish acoustics project (see part 1 of the conversation, footnote 1) is very specific. The software is designed not only to evaluate acoustic data, but initially also to acquire these data. These are two fundamentally different topics. We started with data acquisition for the acoustics project in 2008, with a major breakthrough in 2010. Within this time period, three bachelor's and master's theses were written, dealing only with software development. These theses were not written by biologists, but by computer scientists. For the software we use to operate the Spitsbergen monitoring station (see part 1 of the conversation, footnote 4) as well as the Heligoland monitoring stations we backed a different horse. There we used the sensor manufacturer's original software and made these programs scriptable by fully embedding them in a specific macro-language, allowing us to completely automate user–program interactions. This allows us to basically simulate a person who sits at a computer and presses a certain command at a certain time or upon a certain action, in order to tell the computer to do something: for instance, to start a recording or to load a data file from a sensor. Since the required actions are always the same, it is easy to simulate them with a script. When you see the programs in action, you could thus think that they were being operated by a person. In reality, though, it is the computer that executes the action. This has a major advantage for us. If we take, for example, ADCPs, devices to measure current, these sensors themselves supply only binary data. If we were to write our own recording software, we would have to translate the binary data transmitted by the device into information we can read ourselves—for instance, current speeds in meters per second—which I have no clue how to do. So we take the manufacturer's original software and let it create an output file with binary data every hour. Once we have generated these original data, my computer simulates that I read the dataset into the original program of the sensor manufacturer and translate it into ASCII data. This is a continuous process: The computer receives an original dataset every hour and then begins translating it into data I can read. Ideally, I have nothing more to do with this automated process. I don't get involved again until the readable data are available. From then on, the focus is on the statistical evaluation of the data and on visual representation.

Gabriele Gramelsberger   I have a somewhat different recollection of the collaboration with the FH in Wiesbaden (RheinMain University of Applied Sciences) for the acoustics project.

Philipp Fischer   This collaboration began around 2009. At this time I noticed that I was stuck at the data handling because these data require a really different level of programming. For this project it was necessary to actually write generic programs of our own, since the scientific questions were highly complex and there were no standard programs available to do the required data analysis. Therefore we started to write our own analysis programs in C++ or Java, which I, for one, was not able to do, or at least not at the required level of complexity. This was also the first time that creating a program became a scientific task itself.

Gabriele Gramelsberger   Because the scientific questions were different, or what was the reason for this change?

Philipp Fischer   We simply, and very quickly, noticed that the data flood we were receiving from our acoustics project could no longer be handled with the available software. When we started the first experiments and received the first gigabyte of data, we recognized that we were lost. At this time, coincidentally, a computer science student from the FH

Wiesbaden had just asked whether it might be possible to write his thesis on this topic he found on our homepage. That was the beginning of the collaboration with the FH Wiesbaden and, basically, also the start of the underwater node project.[1] The computer science student who came into our lab ultimately also did all of the programming for the underwater node project, which allowed us to start continuous, real-time data recording with our underwater monitoring sensors. The student, Jakob Klaus-Stöhner, therefore wrote first his bachelor's thesis, and then his master's thesis on acoustics, and later did the majority of the work to develop our monitoring systems, simply because he was able to thanks to his extraordinary skills in computational sciences and software engineering, and also because he had quite a lot of fun doing this challenging work.

Christoph Hoffmann It appears to me that we are encountering two processes here. On the one hand, programming tasks develop from the insights and problems that are acquired during experimentation. On the other, research possibilities emerge from the development of programs. This relationship between experts for software and experts for scientific questions seems central. Who is driving whom?

Philipp Fischer I believe that there is no unequivocal answer to this question. The acoustics project started with the simple question of how we could analyze these data. We had a fast Fourier transform in mind, but we didn't know much more than the concepts involved. Then the computer scientists came and presented us with certain ideas—for instance, localizing fish through the triangulation of acoustic signals. That the project developed in this direction at all is, once again, thanks to the software developed by Jakob Klaus-Stöhner. Six years ago he said to me: My goodness! We have to separate the signals to do that. I said that this would be great if it were possible. He said he had an idea. Subsequently

---

1   As part of the *COSYNA – Coastal Observing System for Northern and Arctic Seas*, the Helmholtz Centre Geesthacht operates three underwater nodes with power and data lines that link sensors and a land-based server station. One of these nodes is part of the observatory in Ny-Ålesund.

he managed, at least in the laboratory experiment, to achieve localization in ten- to fifteen-degree segments. When he showed me that result, we actually did set up new experiments in which the localization of the fish was included as a new method. I never would have had the idea without him. The question—who drives whom?—is entirely justified in view of such internal dynamics. But it is more of a sequence of ideas from different persons. We pose a question, and the excellent computer scientists with whom we collaborate go beyond that question we defined initially and just play around to see what kinds of other things they can do with the data and the experimental setup. That is also true for a different question: When Jakob analyzed fish sounds along with one of our master's students using his new software back in 2013, 2014, he was always really annoyed that we had a maximum recognition rate of about 70 percent. Meaning: When a sound stream was analyzed with the software, it recognized only 60 to 70 percent of the fish sounds that had been heard by a well-trained human ear. We kept checking this over and over, listening for hours and days, and evaluated the data. As a computer scientist, this really challenged Jakob. He and his professor in Wiesbaden deliberated for a long time about how to resolve this problem. They speculated that the kind of tone detection we know from studying vertebrates, mammals and humans, and from the terrestrial domain in general, was entirely wrong here, as we were presuming that we needed to detect oscillations. They tried out a little program based on the hypothesis that the sounds emitted by the fish were not oscillations in the classic sense, like the ones we produce with our vocal cords. Since fish generate sounds by operating a muscle on the air bladder, the two computer scientists considered it from the biological perspective and determined that what we heard was not actually a sound, but a high-frequency drumming. A one-zero signal. The frequencies, let's say 600 Hertz, are not 600-Hertz oscillations, but a 600-Hertz drumming. A Morse signal. We know that the swim bladder muscles of fish are the fastest-contracting muscles in the animal kingdom, reaching over 1,000 Hertz. The contraction frequency would thus allow for the possibility that this signal is emitted. We cannot yet say whether this hypothesis is right or not. So far we are able to formulate the hypothesis that fish sounds are not real sounds at all, but a Morse code language. That would explain, for instance, why we hear those typical knocking sounds, why there are sometimes three knocks and sometimes six, sometimes seven. Until now we simply took it as given that fish sounds are identical to terrestrial vertebrate or mammal sounds. But if we now look at our results so far under the hypothesis that it might be a Morse code, something entirely new emerges; a completely new perspective on fish communication. Perhaps we could also achieve a significantly higher detection rate of the sound emissions from fish in our experiment. We are not yet at the point to decide whether this hypothesis is right or wrong, but if it were to be confirmed, it would be a kind of revolution for fish acoustics in particular, but also for behavioral biology in general. This consideration comes exclusively from computer science. We would not have been able to reveal it in this form at all on the basis of only biological expertise.

Hans-Jörg Rheinberger I have a simple, but it seems to me, not entirely un-important question. Is the term software actually identical to the term program?

Philipp Fischer For me, program specifies software. Typically, I work with the software R or MATLAB. The program I write is written in R. The program is a unit for which I use the programming language of a certain software package.

Hans Hofmann I noticed in our earlier discussions, too, that we rarely use the term software. We talk about scripts, we talk about programs, we talk about coding, we talk about analysis pipelines. When software is talked about in my area, then it is more likely to mean a commercial software package that we, in fact, prefer not to use. We rarely discuss hardware, because most of the people in my area do not think at all about where all of their calculations actually take place. For them that's quite far away.

Christoph Hoffmann If I understand it correctly, the reservations about commercial software have to do with the fact that you always have to seek special solutions?

Hans Hofmann In genomics, and in biostatics as well, there is generally a skeptical attitude toward proprietary software—that is, software that belongs to a company and whose underlying code is not available to the general public—because one never knows what actually happens inside. We do everything with R or Python, and generally with Unix. A well-established software package in computational neuroscience and systems neuroscience is MATLAB, which is proprietary and also relatively expensive. I think that the skepticism toward proprietary software in genomics stems from the fact that many of the methods and approaches are not standardized. Everything is in flux, everything is very new, and the technology changes unbelievably fast. As do the questions that can be addressed with the new technology. Three, four years ago, perhaps even one year ago, the questions that are generating a great deal of interest at the moment could not have been asked in this way. So we need flexibility in practice so that we can invent the appropriate methods of analysis. R has now developed a variety of different packages that allow a broad range of analyses, but are also quite flexible in terms of the size of the datasets and what the data look like. The same is true for Python. These programs, developed by software developers, are already included with the packages. We then apply them by setting our parameters, etc.; everything can be customized. In more complex analyses, these packages are sometimes stitched together into a pipeline. In such cases the data format has to be changed for the transitions, or we have to re-sort the data. For such things you can often use Python. You can even automate this process by writing what are called wrappers. That makes the whole thing even easier. Whenever a standardization starts to emerge, it may well already be obsolete by the following year.

Hans-Jörg Rheinberger I used to be one of those old-fashioned scientists who worked without the aid of a computer. But I imagine that in the end, such programs are sophisticated versions of the protocols on data collection that we were using in the laboratory. They consisted of a sheet of paper stipulating all of the things that had to happen to ensure that the setting was such that data could be generated. Then there was another sheet of paper, also half-standardized, with a protocol on data evaluation. Not quite rigid and fixed, but flexible enough that, as a rule, one could always fill in and process one's data according to the particular context.

Hans Hofmann I would agree: It's the same thing in principle. If we want to extend this analogy: Such a sheet of paper can be quite imprecise. Often you'll see annotations by hand: Then I did this or that. With many of our computer analyses we are practically at the same point. This brings us back to the problem of reproducibility. Everything must be documented: What was actually done, the modifications or innovations, since otherwise it may not be possible for the person who performed the analysis, let alone any other person, to reproduce the analysis. Different results are obtained. This is one of the challenges at the moment. There are platforms like GitHub on which people work with version control and similar approaches, but it's still early days. Some researchers are working hard to take up the challenge; others are less concerned.

Hans-Jörg Rheinberger I want to come back once more to my own laboratory experience. I was doing biochemical kinetics—an experiment with which one can follow the deployment of a biochemical reaction—and for this I pipetted my samples at defined intervals in order to stop the molecular process taking place in the test tube. Then I processed the samples and placed the results in a machine. This machine measured the radioactivity in the samples—that was the marker that allowed me to follow the temporal process. The individual samples were the traces which were generated in the experiment. What came out of the counter, a column of numbers, those were my data. For evaluation I plotted the data according to the Scatchard rule; that is a diagram invented by George Scatchard for testing enzyme activity with respect to cooperativity. For each of my samples I got a point in a Cartesian coordinate system. Then I was able to see whether one or two or three straight lines could be drawn through these points. Depending on the line, one had to decide whether the process that took place in the test tube was a one-, two-, or three-stage process. Now for the computer. I am using a program, let's just call it the Scatchard program. Now all I have to do is enter my numbers in the computer, and the processing, including the graph, is performed by the machine. And now there's a second computer, which controls a robot run by another program. This takes a sample automatically at certain time intervals—now they can be much shorter than the ones that can be taken by hand—and channels them into a sample processing system, also automated, at the end of which the radioactivity counter is located. But now its output is passed on directly to the first computer, which ultimately plots me a graph. Thus, I have delegated my previous activity—except for making the base mix for the reaction in this case—entirely to the machine. That is, I think, what Gabriele meant at the very beginning of our conversation, with the formulation that "the observer is replaced"—only, I would say that the experimenter is replaced. But is that everything, so do we have a complete analogy, or where does it get exciting? Computer one can, of course, calculate more precisely, and computer two, which controls the robot, can pipette faster and supply more samples and thus more data to computer one. Therefore, I imagine, a considerably altered graph may ultimately result, which can inspire me to a new idea regarding the observed molecular mechanism. But in the end, I have to look at the graph myself and decide whether or not it makes sense.

Philipp Fischer I really like the analogy with the sheet of paper. When we do an experiment and the quantity of data is manageable, I do not allow my students, much to their dismay, to use the computer. If we have a single factor ANOVA (Analysis of Variance: statistical test for multifactorial experiments) or a chi-square test (statistical test for frequency distributions), they have to do the calculations by hand. Then I am sure that they actually understand the calculations. The computer, the program, R, whatever, only actually has a role when the data volume becomes so large that it no longer makes sense to do the calculations by hand.

Hans-Jörg Rheinberger So it is a question of quantity rather than quality?

Philipp Fischer It is a question of quantity. To a limited extent, yes, there are analyses which we can hardly calculate, but ultimately one can trace it back to the quantity.

Hans Hofmann The data are perhaps more complex, but I would agree, the principle remains the same.

Software Cultures and their Agency

Hannes Rickli Is that a difference from other areas? At CERN, physics, software and programmers play a very different role. In biology that would probably be restricted to the context of sequencing. Martina Merz investigated this software agency.[2] You probably know more about this. What's the situation in meteorology?

Gabriele Gramelsberger It's quite specific to each given discipline. In meteorology you would never use MATLAB. If you use MATLAB, you're out of the game. You have to program everything yourself in C or in Fortran. In contrast, MATLAB, as far as I know, is the standard for biologists.

Hans Hofmann Not in genomics or computational neuroscience, where the younger folks mostly prefer Python. That also has a lot to do with the fact that MATLAB is simply too expensive. Many universities are unwilling, or no longer can afford, to purchase a site license; and a per seat license costs a lot of money, especially for a small lab. However, in other areas of biology MATLAB is still quite strongly represented.

Gabriele Gramelsberger I would call it software cultures. There are different cultures in different disciplines, and even in the different labs.

Christoph Hoffmann Could these differences have something to do with the fact that in some contexts the point is to resolve problems—so to speak, menial work—whereas in other contexts software development and writing programs are a component of the actual research process?

Philipp Fischer That's what I meant earlier. In the fish acoustics project we had some phases when our computer scientists told us what kind

2 Martina Merz, "Kontrolle – Widerstand – Ermächtigung: Wie Simulationssoftware Physiker konfiguriert," in *Können Maschinen handeln? Soziologische Beiträge zum Verhältnis von Mensch und Technik*, ed. Werner Rammert and Ingo Schulz-Schaeffer (Frankfurt am Main: Campus Verlag, 2002): pp. 267–290. See as well Martina Merz, "Multiplex and Unfolding: Computer Simulation in Particle Physics," *Science in Context* 12 (1999): pp. 293–316.

of experiments we could do with this and that software. In this case the software people did not solve our problems, but instead in effect prescribed the scope of action in which we were able to conduct our experiments.

Gabriele Gramelsberger   That is an extremely good example, but I imagine it is rarely the case. Software-driven research—that would be a terrific thing.

Hans Hofmann   In bioinformatics, computational biology, there is certainly a tension between the people whose career and program of work consists in developing algorithms and analysis methods, and those who pose biological questions and apply these methods. More and more people from the second group are of the opinion that the members of the first group deserve recognition. But this is not institutionalized everywhere when it comes to promotion and tenure and the like. Those in the first group may perhaps have a greater problem, because their publications tend to be collaborative. In a conventional tenure track it is then often not clear which authors contributed what to the paper. This is starting to change, and some universities are further along than others, but like everything in academia it is a slow process. In any case there are still colleagues who see the development of programs as a subordinate activity.

Christoph Hoffmann   In the interviews at the Center for Computational Biology and Bioinformatics in Austin I presumed that the advisors working there come into play when the datasets are collected and it's time for the analysis. In conversation I learned, however, that their work starts with the design of an experiment. For example, the entire experimental procedure is oriented toward the generation of a certain type of data that can be processed well. That astonished me. Their contribution is indeed much greater than that of a "mere" technician.

Hans Hofmann   That is true in any case. Here a somewhat different perspective comes into play. The consulting team, our staff scientists, want people to come to them as soon as they think of an experiment for the first time. If they come to us only when they already have the data, we often have to conclude: The questions they have about the data cannot be posed at all because the experiment was not conducted or planned in a suitable way. Then we say: "Bad data in, bad data out." Even with the best bioinformatics, there's nothing you can do.

Hans-Jörg Rheinberger   Those are all examples that concern quite concrete questions. It just occurred to me: Many years ago—I think it was fifteen years ago—I interviewed Gerhard Kremer, the former representative of the Packard Instrument Company in Zurich. Packard was a pioneer in the development of scintillation counters, a technology for measuring radioactivity, important for counting labeled molecules. In the 1960s and 1970s this device was found in many molecular biology labs. Kremer spoke of this technology as an enabling technology.[3] Thus a technology that can be used in many ways and makes new research approaches possible. The question would now be whether that can also be said of software. Can one view the development of software as a relatively autonomous matter that has the character of an enabling technology? People from different laboratory contexts can access it and then adapt it according to their needs.

Philipp Fischer   This becomes more and more the case. Today, software can significantly accelerate research. Procedures and evaluations are faster, and I can make progress in my research much faster compared to earlier times. But a great deal of discipline is necessary to keep from getting lost in the technology.

Hans Hofmann   I think so too, definitely. There are such heroic figures in both R and also Python. Hadley Wickham is a good example, or the people who are deeply involved in the R Bioconductor project, and it's similar for Python. They write these packages and develop them further in order to offer users like me ever more possibilities. Hadley Wickham is renowned in the scene for his great package ggplot. It gives you an unbelievable amount of flexibility to make figures of all different kinds. Other

3   See Hans-Jörg Rheinberger, "The Liquid Scintillation Counter: Traces of Radioactivity," in *An Epistemology of the Concrete* (Durham: Duke University Press, 2010): pp. 170–202, 171.

people have written other packages that are very influential and have won over quite a fan base. When Hadley Wickham comes to Austin—he's from New Zealand and has been with Bioconductor since nearly the very beginning, as a student; Bioconductor is the nonprofit organization that drives R code forward and distributes it around the whole world—all of our students say: "Oh, Hadley is coming, we have to go and see him!" No matter what he decides to talk about. It's a relatively small group of people who write these packages, but they have a great influence both on the enabling technologies and on how we render our data and which possibilities we take advantage of. After all, the software was born in their brains. But only in recent years has anyone learned the names of these people. That is a generation gap. Many people of my age or older generations don't know them, whereas the grad students, the postdocs, they all do.

Philipp Fischer This is primarily because we have been working more intensively with software like R in recent years, which is open source. People cite and address the R package, ggplot and many other open source software packages like a sort of scientific publication. Manuals and instructions for software packages are published in the same way as I publish a scientific manuscript within which I summarize my knowledge. I think that the development of open-source software is a very significant point leading to recognition for this activity, so that it can be credited as well.

Gabriele Gramelsberger I think there are many standard algorithms and methods which are often named for their inventors. Like the Klett algorithm, for instance, which is used to collect data with LIDAR (light detection and ranging). The algorithm specifies a certain inversion ratio in the LIDAR equation so that measurement data can be acquired at all. In 1981 James Klett proposed this first algorithm to resolve that problem.[4]

4   James D. Klett, "Stable Analytical Inversion Solution for Processing Lidar Returns," *Applied Optics* 20 (1981): pp. 211–220.

Understanding Software

Christoph Hoffmann For me this raises the question of how this can be researched. How can I follow the role of programs in the research process, how can I write their history? You are attempting to write this history with the help of version control. But my goal is not to document version after version, but rather to understand what happened between the versions. This then concerns things that are not recorded. If I have something material in mind, perhaps a certain experimental arrangement like the fish acoustics project on Heligoland, I can construe how individual changes will affect the experiment. But for a software package or a program you write yourself, it is obviously not always easy to recognize exactly what it does with the data.

Hans-Jörg Rheinberger This is always unbelievably context-dependent. I want to come back once more to my own time in the laboratory. At our institute, we had a subgroup that studied macromolecular complexes by way of neutron scattering. Ribosomes were their subject. There were three environments. One was biochemical: A macromolecular complex had to be biochemically isolated, and then individual components had to be obtained from it. Some components were derived from deuterated bacteria (bacteria grown in heavy water) and others from bacteria that were washed in normal water. Deuterated components in a non-deuterated environment: This was a way to measure distances between the components. The second environment, the neutron spallation source in Grenoble, was located 1,000 kilometers away. You had to deep-freeze your biochemical probes, and then fly there when you were granted measurement time. The neutron scattering data were collected there. The third environment was, finally, the data evaluation. And that was software. But the software was applied in an extremely targeted manner: It served to determine distances from scattering data. A relatively clearly defined task; only mathematicians can do it. The biochemists on the other end did not have to know much about it. But everyone knew that, in principle, the software could not be better than, first, the biochemistry that supplied the original input, and second, the generation of the scattering data. Thus, I think you get an idea of how embedded in the process of experimentation things like software really are. Indeed, it is important not to hypostatize the software as an entity that starts to lead a life of its own.

Christoph Hoffmann It seems to me, though, that this instrument is, on the one hand, softer than a physical instrument, in the sense that it is constantly in flux. Is there such a thing as a fixed program? Or will yet another component be added tomorrow? On the other hand, I have the impression that the way this instrument works makes it much more complex than something like a centrifuge.

Hans-Jörg Rheinberger That I don't believe.

Hans Hofmann The centrifuge is very complicated, too, and most people who use it in the laboratory cannot tell you how it works or how it is constructed. But one would certainly wish that when people use a program, they understand how the data are transformed and what happens with the data. But they do not necessarily need to know how a program is written. They do not have to have that expertise.

Gabriele Gramelsberger In computer science they call this the shift from programming to tooling. It is problematic. Of course, you must be an informed user and know which tools you are plugging together and what effects this has, but that is, of course, a different situation than having to program yourself. I know this from other fields. At the moment there is a discussion about whether what is happening there is a fundamental shift, a generational shift.

Hans-Jörg Rheinberger In principle that is the case with nearly every research technology. Two generations ago, the only people who could work with ultracentrifuges were those who had the corresponding know-how and were able to build parts of the machine themselves. But this was true only until they were black-boxed. There are very, very few research technologies that have resisted black-boxing for any substantial period of time. Electron microscopy is one of them; you still have to have experts to work with such an instrument.

Christoph Hoffmann But we already established that there are different software cultures. Philipp tells us that he always writes his own programs for data evaluation. Is that the case for your students as well? Do they write their programs themselves, or do they, as Gabriele described it, plug components together? Is the process becoming ever more of a technical routine, or is it still a handicraft?

Philipp Fischer On the contrary, I believe that we are currently experiencing a turnaround. Five to ten years ago our students were exclusively users of software. Today they (at least my students) are learning to program themselves again. What Hans also said before: When people finish at our institute, I expect those who have written a master's thesis, and everyone who has written a doctoral thesis, to be able to program their own analysis routines in R or MATLAB or whatever software is appropriate. I do not expect them to learn Fortran or C or Assembler, but they must be able to use at least those programming languages which we use in biology today.

Hans Hofmann The students and scientists in general should be able to develop their own analysis approaches. They have to understand the data structure, formulate their questions clearly, and visualize their results in a convincing way.

Philipp Fischer I want to take a moment to tell a brief story about something that just occurred to me. The story shocked me, but it also shows a conflict area. A while ago I was talking with a student who was about to submit his doctoral thesis in medicine. The following astonishing conversation began to develop when he told me: "Hmm, so I'm finished with my data collection, now I gave the data to the statistician and I think I'll get them back in three weeks." "What, you gave your data to a statistician?" "Yes, he has them now and is evaluating them. When I have the results I'll compile them and finish my dissertation." This left me, honestly, quite stunned. That is an anecdote, which, I believe, shows the difference between users and those whom we call scientists.

Christoph Hoffmann So is it ultimately about control?

Hans Hofmann Not only about control, but about being at all able to ask the right questions. I would call it ownership.

Eloquent Data

Philipp Fischer I apply statistics, of course, with a certain conception of what I want to do with my data. But in the evaluation process I also do learn a lot about the data. I become much more familiar with the data variance, how the data "behave" in analysis and what their dynamics are, when I work with the data intensively. In the process of evaluation, in the—I'd like to exaggerate a bit here—communication between me and the data, I suddenly get ideas: Take a look at this or that.

Gabriele Gramelsberger Let the data speak.

Philipp Fischer What does it look like when the data tell me, hmm, could that really be? Could I think in that direction? Do the data say, do the data perhaps enable me, to work further in this direction? I cannot do that with any standard program.

Hans Hofmann Philipp already said that as well. Of paramount importance are what we call sanity checks. You have to look at your original data exactly from all possible angles to assess: Are they plausible, does that make sense? The purpose is to exclude the possibility that something or other went wrong, and to ensure that the quality does in fact meet your demands, before you do any standard analyses or other kinds of processing. If this is not done, and you give them to some statistician—who knows absolutely nothing about the quality problems that can be associated with the data—you may perhaps end up with wonderful results that have nothing to do with reality. Sometimes such results are even published, and when the quality problems come out, we think: How did that get in there? Was nobody paying attention? I do not want to be in this kind of situation, and I do not wish my students to find themselves in such a situation, either.

Gabriele Gramelsberger   That speaks for an extreme contextualization of data. Data cannot simply be uncoupled and transported from one place to another. Data are not something I just hand over to a statistician, for whom it does not matter whence the data come, because he/she always does the same statistics.

Hans-Jörg Rheinberger   I think it is important, as Philipp is saying, to conceive of data processing as a creative process. It is not simply a routine, and it never falls into a habitual routine. You learn just as much about an experiment during data evaluation as you do during the process of generating the data.

Hannes Rickli   Also important is the communicative aspect that Philipp mentioned. Apparently, the data speak as well. You have to set your sense of hearing such that they can convey something you may not have known before. The risk of handing data to a statistician is that he evaluates them in a quasi-automated way and does not recognize particularities that have not been seen before.

Hans-Jörg Rheinberger   For what is called hypothesis-driven science, it is clear, and to some extent trivial, that data are collected with a certain intention, and that processing can then take place with this perspective in mind and not any other. Functional data are not structural data, for instance, although structural data can, of course, help interpret functions. But as Philipp said before: The processing, the manipulation of the data can expose new clues, just as can data collection. The data-first people say: Let us collect data first, the questions will come up afterward, during processing. That would be the extreme form. Data collection then belongs to the infrastructure, so to speak, it is purely technical and instrumental, and science in the proper sense of the word is then performed in the data space. In this connection I would ask: Is the differentiation between data generation and data processing, which we have been using so often here, something fundamental? Can one be decoupled from the other, or in what way must generation and processing be related to each other?

Hans Hofmann   With a few exceptions, practically all of the projects in my laboratory require experiments and then data analyses on big data, bio-informatics and the like. For the students and postdocs who conduct these projects, this is a quite natural, iterative process. The first thing they do, for instance, is various analyses on different samples. This then tells them how the remaining samples must be processed. It always goes back and forth. For this generation it is a completely natural process. What's very important—I already mentioned this before—is ownership: that you can actually regard the entire project, the experimental part and the data analysis part, as your own. Our institution is most of the way there by now. I don't really see this decoupling anymore.

Philipp Fischer   I have another example in which statistical evaluation already intervenes in data collection. We discuss this with our students over and over again, with moderate success: that it is highly advisable to start an experiment—as every statistician knows—with statistics, before a single datum is recorded. We always try to teach our students the advantages of power analysis. With a certain statistical procedure I can thus determine beforehand how many replications I have to perform with how many test organisms in order to prove that a certain result is statistically significant. That means I can determine how much power this experiment has. This is a classic example for how statistical methods can serve as a framework for an experiment. When we work with fish, that is, when we work with vertebrates, we have to apply for a legal permit to do experiments with them. We always underpin these research applications with a power analysis so that we can provide statistical evidence: If we kill 20 animals, there is a 95-percent probability that we will be able to discover a difference between two given treatments. Our ethics commission asks for exactly this proof. Not, for instance, that we take only ten animals and then have to say, at the end of the experiment, that we should have taken 20 in order to be able to discover any existing difference between the treatments. Then the ethics commission would justifiably say: "That is all very well, Mr. Fischer, now you have killed ten animals, but according to your statement, you cannot say whether there is a statistically significant difference between these two treatment groups. The power of your findings is only 50 percent. You should have taken 20 animals. Then we could have said that the result you are presenting to us is correct. Because you took only ten animals, they died for naught." This is just an example of how important the integral concept is. In my eyes, of course, data collection and data processing are two subsequent steps. But ideally, there can't be one without the other.

Christoph Hoffmann I would like to come back once more to the eloquent data. Before this conversation, Gabriele once raised the question as to how anything new can be discovered when the program so restrictively narrows what can be focused on. Is a program perhaps not open enough for surprises? Philipp, you said that the data speak to you, but they speak to you through the program you wrote. There is something in between. How do you see this?

Philipp Fischer I believe that we have to redefine the activity of programming. It is not as if I write a program, sit in isolation for four days and then apply this program to data. I start writing a program and make a test version first. I run my data through it once and see: Hmm, the evaluation is perhaps not quite as clear as I'd like. So I refine and optimize this analysis here and there, and take a look at the results of the different analysis steps I just added, and say: Hmm, okay, that's not bad at all, but, hmm, look here, if you maybe divide the subgroups again, or if you apply a certain analysis method there, what comes out then? Programming is an explorative act. In the ideal case, I already thought about exactly what I will do with the data before I start any data analysis. In fact, I have to. Once I actually have the data in front of me, my preconceived evaluation strategy is applied, but I nevertheless set my mind to looking at what else I can get out of these data. After all, this dance with the data allows for various perspectives. It is not a fixed, rigid program.

Hans Hofmann That is what I meant before when I called it iterative. What Philipp is saying is definitely true. And then we also have these analysis pipelines, which are relatively standardized. One may perhaps have to configure other settings, etc., and then this raises the question: Can that now in any way reduce the possibility of discovering something new? However, each experiment is new; you ask completely different biological questions. They are entirely different experiments. You produce only data that have a certain format, which are similar, let's say sequencing data, and due to conventions and certain necessities they must be processed in a certain standardized way. But what ultimately comes out of it cannot be foreseen and allows you to make new discoveries. When one has sequencing data, when one has ensured that gene expression, or differences in gene expression, can be measured, then comes the next step. That is when I say: Okay, I have all of these different experimental conditions, I have all of these diverse data for different genes, etc. How do I visualize that, what kind of multivariate analyses can I do? That is, again, a very creative process. All of these things happen simultaneously and iteratively in any project.

The Incompleteness of Software

Gabriele Gramelsberger The question "what is new" was directed to the software in the fish acoustics project which decides: that is an event—a possible fish sound—or not. The things that are classified as neither an event nor a nonevent, but rather as unclear, are the ones you can make discoveries with, about which no automated decision is made. I conducted interviews with programmers in meteorology.[5] The most productive thing for them was always the error productivity. You program something, you get data, it doesn't work. Now, of course, I am talking about simulation models. They spent ages troubleshooting, because the simulations did not agree with the real data. This is where they learned the most about their program and about the theory behind it that had been plugged into the model. That was iterative: One runs the model again, calculates, data come out, one compares them with the measurement data again, and so on.

Hans-Jörg Rheinberger So in this sense software is something that is never completed?

---

5   Gabriele Gramelsberger, *Computerexperimente: Zum Wandel der Wissenschaft im Zeitalter des Computers* (Bielefeld: Transcript, 2010): pp. 171–176.

Gabriele Gramelsberger No, never.

Christoph Hoffmann But that is also the problem. Are commercial software and individually adapted programs really instruments in the sense of something technically stable and reproductive, or are they something between instrument and research object?

Gabriele Gramelsberger Software is soft.

Hans Hofmann It is an instrument in the sense of music. That's never completed either.

Hannes Rickli The problem of reproducibility comes up here, too. That's why we have the snapshots to document each version of the program.

Hans-Jörg Rheinberger Software without a certain degree of stability is no longer software.

Hans Hofmann In the end, the community and the convention say what level of stability is acceptable. When a new technology becomes available, the standards and expectations are often somewhat lower than later on, when it is easier to understand what can go wrong and why.

Gabriele Gramelsberger From meteorology I know that there are version cycles. Many different people work to develop a software, then there are deadlines when the entire community meets to produce a software release (Version X.0). The next day everyone starts again to develop the software further, and so on.

Hans Hofmann That is also important because the supercomputers, the high-performance computing systems, computing centers, whatever you want to call them, have to perform regular upgrades and updates. That generally happens only when a new version is available, because it entails time and expense, and generally causes all kinds of minor problems. When you go from 2.2. to 2.3 and there is no major difference, they simply skip it and say they'll do the next update when 2.5 is out. This varies depending on the software. Some software packages have a good reputation, they are stable, they do not cause any problems in the environment in which they are embedded. Others do not have such a good reputation. Every time they do an update, something unforeseen happens. In the end the people decide; it's a matter of supply and demand. If something is good, robust, continues to be developed, and has good documentation, it is implemented and continues to be improved. If that is not the case, the software lands in the dustbin of history; that happens quite fast.

Christoph Hoffmann We have not only data dumps, but also program dumps?

Hans Hofmann   You should have a look at what all is on my server, there are script dumps from the last 15 years.

Gabriele Gramelsberger   Cemeteries.

Software-Based Experimentation

Gabriele Gramelsberger I have a question. Would you say that the "experimentation with" is shifting into the software? Do you use software for experimenting? We had a couple of examples for which that could be assumed. This would expand the concept of experimentation.

Hans Hofmann   Yes. You do that, for instance, when you have a complex dataset. You use different statistical models, different algorithms, different programs, to explore the data, to view them in different ways, and you try out different ideas. I definitely see that as close to experimentation. For younger researchers there is hardly any conceptual difference from experimentation in the wet lab. This is not always the case, but at least the more complex projects—in which large amounts of data are generated, in which you have many variables and need multivariate approaches, etc.—are definitely experimental. Also, statistical approaches have become increasingly important that allow you to determine the extent to which different statistical models explain your data. How do you determine which one is best? Or is there no such thing as a single best one which you then publish? By now it has become apparent that this may not be the right approach at all. Instead, one takes these 20, 30 models and develops a criterion, often called an information criterion: One attempts to calculate for a model how much information content it has relative to the data. Building on this, you can create a ranking and develop criteria that define the point at which there is a cutoff, and which model you find most credible, which you can then interpret further. Or several different models are selected. In certain areas of biology this has become routine. In this sense there is not only one truth. That is a very experimental approach, for you may potentially test 100 different hypotheses or more at the same time.

Philipp Fischer   One thing that occurs to me where software does in fact serve experimentation is, once more, fish acoustics. Here we have already conducted identical experiments with the same animals on several occasions, but using different software. We wanted to find out which software is best for the experimental result. In this case the experiment was oriented completely toward the software: We had three different versions, and it was not entirely conclusive which method was best, so we performed the experiments three times.

Hans Hofmann   You can take that even further. We have often conducted simulation experiments in which you use simulated data and compare different programs to see which has the best performance.

Gabriele Gramelsberger   Almost a kind of benchmarking method.

Hans Hofmann   When you develop programs it is in your own interest to do benchmarking. That way you can convince people that your program performs best, or very well, at least under certain conditions. In practice, this has become the standard by now and is also experimental in the same sense.

Hannes Rickli   I am interested in what's "better." From the artistic media I know, of course, that the settings of a filter or a program expose entirely different layers of signals. When I talk about what's better, this is oriented on my presumption. The question is: Does this presumption change, this hypothesis change, through testing the software? Or is the software, in the end, simply the best because it is most likely to confirm my presumption? It could also be that certain software reveals something that was not intended in the framework of the presumptions.

Hans-Jörg Rheinberger I believe that these two go more or less hand in hand. Normally you operate in optimization mode. You want to manage things as best as possible. But if you have enough sensitivity to the disturbances, they can point out a new lead. You do not just start with an experiment and claim that you are completely ignorant. Generally, there is something you want to optimize. Or at least you have an idea about what ...

Hans Hofmann ... is supposed to happen. You just have to be open to the possibility that what comes out is not what you expected, that something else may happen.

Hannes Rickli And this something else could definitely become interesting. If one doesn't have it on the horizon. Here again, this shift. The medium as an epistemic object.

Hans Hofmann That is quite normal. I can give you many examples of something surprising coming out in an analysis, which then completely changed the next steps. Just as they do in wet lab science. The same thing happens there. The general public often does not understand this. By definition the experiments must be going wrong, because, after all, we are doing something that nobody has done before, and therefore we often cannot predict what exactly will happen. Or you start with a project, you have certain presumptions, you have certain hypotheses or ideas, and at the end you are somewhere completely different. But that's what makes it so exciting. If I always know beforehand what the result will be, I am not a scientist. Then I am ...

Gabriele Gramelsberger   ... an engineer.

Hans Hofmann   No, that doesn't make me an engineer, either. In English I differentiate between research and science. When I do science, I do not necessarily know beforehand what the results will be. I always have to be ready for surprises. When I do research, for instance, test toothpaste on some cell cultures in order to fulfill regulations, I have a relatively clear idea what will come out of it. I simply check off one box after the other. Then I am doing research, but I am not doing science, because I am not discovering anything new about nature.

Hans-Jörg Rheinberger   That is an interesting use of the terms. In our circles one would turn it the other way around: Research is accompanied by uncertainty, while science is what is established.

Hans Hofmann   In any case, for me data analysis, or data exploration and the like, have become part of experimentation.

Hans-Jörg Rheinberger   But that is indeed the range of experimentation. On the one hand, you have an explorative pole, and on the other you have the pole of testing. Between these two extremes there is an unbelievably broad intermediate scale.

Christoph Hoffmann   Now a strong emphasis has been placed on the expansion of the experimental space into the space of data analysis. But viewed historically, back once again to the sheets, this is actually nothing new. The difference between what was once done on paper and what is done on the computer today is primarily the medium. Today we experience merely that the analysis space has become more complex. The possibilities for dealing with data are becoming more diverse, time plays a role.

Hans Hofmann   It is interesting that you go back to history. I can no longer remember the exact quote, but even 200 years ago there was the concern that too many data were there and one no longer knew what one should do with them. A new method is invented, data can be collected, and the people can barely keep up with writing everything down and placing them in tables. This often led to mathematics being propelled forward, to statistics being invented, these were ultimately data challenges.

Gabriele Gramelsberger But there is a difference, of course: Only with the computer can you do multivariate statistics.

Hans Hofmann But look at the history of astronomy, for instance: At Harvard Observatory, in the decades before and after 1900, they photographed the entire sky in order to see minor shifts that allowed them to discover new stars or phenomena. For evaluation they had a whole army, made up mostly of women with a bachelor's degree in mathematics, who performed gigantic computations as if they were on an assembly line.[6] Each person did a certain calculation step, over and over again. At the end they came up with calculations that people today would say can be done only with a computer. They actually did it all by hand, 20 women in one room, who calculated all day long. That is unbelievable.

Gabriele Gramelsberger But all the same, they computed only two variables.

Hans Hofmann Yes, but the principle is the same.

Gabriele Gramelsberger I'm just saying, that is the cognitive break.

Hans Hofmann They had a data challenge, they had gigantic masses of data that were acquired from the telescope images, and actually per-

6   See Jenny Woodmann, "The Womens' Computer who Revolutionized Astronomy," *The Atlantic*, December 2, 2016, https://www.theatlantic.com/science/archive/2016/12/the-women-computers-who-measured-the-stars/509231/. For more information, see Dava Sobel, *The Glass Universe: How the Ladies of the Harvard Observatory Took the Measure of the Stars* (New York: Viking, 2016). In addition, see Natasha Geiling, "The Women who Mapped the Universe and Still Couldn't Get any Respect," *Smithsonian.com*, September 18, 2013, https://www.smithsonianmag.com/history/the-women-who-mapped-the-universe-and-still-couldnt-get-any-respect-9287444/. The film *Hidden Figures* (2016) deals with the work of female mathematicians for NASA during the space race in the 1960s.

formed these calculations. In the end someone took a couple of numbers from the results and compared them over time: Recorded three weeks ago, two weeks ago, last week, and last night. They showed a phenomenon that was interpreted over the year and ultimately understood. Without these data analysis sweatshops this would have been impossible. We may have different approaches and another technology—and of course, the scale of the data has increased considerably—but when we are so fond of saying that this is a new problem that never before existed in the history of science, we are deceiving ourselves. If you go back 80 years, or 200 years, depending on the discipline and in which context, and were able to talk with the people then, you would probably hear something similar.

Gabriele Gramelsberger I am not entirely *d'accord*. These stories are certainly familiar, and they have been told very well. Nevertheless there is this cognitive limit. For the multivariate calculations you need computers, and this opens up a data space for which, in turn, this comprehensive analysis is needed.

Hans Hofmann And the problems with visualization ...

Gabriele Gramelsberger ... because you can still depict the relationship between two variables just fine with tables and graphs. This breach, that is what gets the whole brew bubbling.

Hans-Jörg Rheinberger But in mathematics, of course, we have been able to think in n dimensions since the 19th century.

Gabriele Gramelsberger To think, but not to calculate—Henri Poincaré made that clear back in 1890.[7]

Hans-Jörg Rheinberger Over the last 200 years, the research technology interstice—as I will call it now—which has edged in between us and the

---

7 Henri Poincaré, "Sur le problème des trois corps et les équations de la dynamique," *Acta Mathematica* 13 (1890): pp. 1–270.

objects, has increased in complexity to an ever greater degree. From the historical perspective, one can perhaps distinguish three phases: The first was mechanics, the second electronics, the third is informatics. One acquires knowledge about nature by implementing ever more techniques, in order to find out something about the objects out there. We ultimately did not answer the question as to whether the informatics interstice has a qualitatively new dimension as compared to the previous ones, or whether it simply makes everything more complex. Gabriele tends toward the former view. What Philipp and Hans describe could be interpreted from a rather traditional perspective as an expansion of options. Do we have to leave the question open?

Gabriele Gramelsberger The way Hans and Philipp even talk about library tools, programs and software! Everyone from the outside, or anyone who has had nothing to do with these things for 15 years, is truly flabbergasted when they see the kinds of tools there are out there today, and the squadrons of them available.

Hans Hofmann You have to stay on the ball practically continuously, otherwise you lose touch in no time at all.

Hans-Jörg Rheinberger For me, I can only say once more, when I worked in the lab 30 years ago, I was able to do good research and did not need any of it.

Hans Hofmann But to that I must say: A good, elegant, simple experiment always has an important place in science.

Hans-Jörg Rheinberger For the statistics I always had to do my double determinations and then divide by two. And there was no shortage of interesting results.

Hans Hofmann Then you probably determined a coefficient of variation, and you were done.

Philipp Fischer Those were great times, when I used to take the boat out on Lake Constance, throw out a net at sunrise and then catch 20 fish. And on the other shore I caught 40 fish. I had my data in my hands when I came back with the boat into harbor. When the data work nowadays really overwhelms me, sometimes I look back fondly on those days.

Hannes Rickli Fishing can, of course, also be understood symbolically, fishing in troubled waters, but then you had something in your hands.

Hans-Jörg Rheinberger Does one not speak of data fishing?

Hans Hofmann Fishing expedition, if it is exploratory. If you're negatively disposed to data exploration, you call it a fishing expedition.

Infrastructures

Materialities and Infrastructures

Hannes Rickli The next part of our conversation concerns materialities and infrastructures as a subsection of materiality. Electricity is a part of the problem of infrastructures. Perhaps I should briefly say why this is of interest to me. Our conversation took a very nice course: Via the data we came to the results of research work. Then we talked about the software and the programs with which data are collected, processed and analyzed. Now we are going back to the base, for instance, to the electricity that is needed, or to the environmental activity that affects the facility in Spitsbergen: the saltwater, temperature changes, and so on. The digital processes are based on electricity. Without electricity, the remote sensing work done by Philipp in Spitsbergen would not be possible. And additionally, of course, the distribution of his data over the internet would be impossible. When we talked about software and programs before, it became clear that either commercially available software is adapted, or a whole new software package is developed. So there is some space we can fill ourselves. The topics we are discussing now, however, are completely beyond our control. Provision of electricity, that is, the availability of energy resources, or submarine cables, connection to the internet, are no longer in the hands of the research enterprise. The question is: Do these elements contribute at all to shaping research work, and how does this emerge in the context of experimental work? Are infrastructures and nature considered factors of uncertainty, or do these peculiarities get lost in the normality of research work? Before we go any further, I would like to show you the latest status of our audio panorama concerning Hans's research on cichlids in Austin.[1] This is still a work in progress, and we are still

---

1   Cichlid #3, Soundscape Texas (Aug. 21, 2014), https://computersignale.zhdk.ch/en/data/cichlid.

developing ways for the recordings to be rematerialized as a work of art. In the middle of the screen you see eight stations. It starts on the small scale with the aquarium, followed by the minus–80-degree freezer in which the tissue samples are stored; this situation represents the molecular biology laboratory, the wet lab. The further stations then belong to the dry lab: the GSAF (Genomic Sequencing and Analysis Facility), where the Illumina is located, the gene sequencer with which the DNA/RNA dissolved in a liquid is digitized. The "digital raw material" is recorded by a server, and then the data material is passed on to the *Stampede* supercomputer at the TACC (Texas Advanced Computing Center) along with the corresponding programs. From this point on, the massive infrastructures take over, the chilling station for the coolant, the electricity plant and a fracking drill tower that supplies natural gas for the electricity plant on the UT Austin campus. This drill tower belongs to the UT lands, meaning, to the properties that the state of Texas gave the university to manage. With this work we want to show how the energetic conditions of the data processes in the dry lab are connected with those in the wet lab, with the freezer and with the aquarium.

Hans Hofmann The whole thing takes 24 hours; everything starts in the dark.

Hannes Rickli At the moment only one station at a time can be viewed in detail. The idea, however, is to show all stations simultaneously as a panorama in an exhibition room.

Hans-Jörg Rheinberger Can you give us an idea of what this panorama would look like?

Hannes Rickli Well, that's the question we're working on. One concept would be a website on which the different stations are shown as lines, one below the other. Then one could move the cursor along the timeline to follow the changes at all sites at the same time. At the moment this only works for one station.

Christoph Hoffmann An essential point is that the materiality should become tangible through the sound?

Hannes Rickli The premise is that digital processes, too, are ultimately physically based and leave corresponding traces. As users, we do not perceive these physical aspects while we are focusing on the operations of the algorithms. They intrude on our perception only when malfunctions occur, if at all. Then these physical qualities come into play, such as when a hard drive crashes, or the ventilation starts buzzing. From an artistic perspective this moment is especially interesting. The biological sciences take place along so many digital processes that one could get the impression that science is, in certain respects, immaterial. When you see Philipp's graphics, the last thing you think about is the difficulties involved in creating them. This is just the same for Hans in Austin: There, too, it is difficult to imagine the connection between electricity and the results of digital data technology. My interest as an artist is to turn this immateriality around and direct the focus to the material side of seemingly incorporeal data. In so doing I place the sonic at the center, because digital processes are primarily time based: They concern temporal microdifferences. While a processor works, energy consumption, for instance, oscillates—depending on what is processed. If the oscillations are recorded by sensors as sequences of values, they can be played back as an acoustic signal. With this, a small excerpt of the physical phenomena can be heard as a sonic phenomenon.[2] Yet some of the signals are located in the transonic range.[3] Sound is the ideal medium, because what we have

2   The sonic is its own category between acoustics and music and concerns sound that is generated not by a resonator, but through technical processes. See Wolfgang Ernst, *Sonic Time Machines: Explicit Sound, Sirenic Voices, and Implicit Sonicity* (Amsterdam: Amsterdam University Press, 2016), p. 7 and pp. 23–24. Here an audible sound is considered from a psycho-physiological perspective. There may not be any semantic attribution as in music, but a cultural conditioning does take place through the structure and functioning of the technical processes. The concept was developed at the Institute of Music and Media Studies at Humboldt University, Berlin. See Peter Wicke, "Das Sonische in der Musik," *Pop-Scriptum* 10, Das Sonische – Sounds zwischen Akustik und Ästhetik (2008), https://www2. hu-berlin.de/fpm/popscrip/themen/pst10/pst10_wicke.htm.

3   "In this sense I suggest the use of the term trans-sonic, which denotes the realm of all signals, oscillations, rhythms and flickerings that are inaudible for humans, but can be made— to a certain degree with certain losses—audible again by the use of media technologies such as audio amplification, radio demodulation or software sonification." Shintaro Miyazaki, "Algorhythmics: Understanding Micro-Temporality in Computational Cultures," *Computational Culture. A Journal of Software Studies* 2 (2012), http://computationalculture.net/ algorhythmics-understanding-micro-temporality-in-computational-cultures/.

here are oscillations in temporal sequence. This contrasts with images, for although these, too, may register light events that are ultimately based on oscillations, they can be fixed in the register of simultaneity.

Christoph Hoffmann   Another way to make materiality accessible would be the temperature.

Hannes Rickli   Right. Temperature oscillations, or even optical signals. The LEDs of electronic components, for instance, flash in a specific rhythm while a computer runs, for instance, to work through an algorithm. Such physical tracks are called traces or side channels in computer engineering, and their investigation side channel analysis.[4] They are what no longer appears in the result—that is, in a digital datum. However, they are the foundation, the very condition for a computational process to take place at all, and thus the occurrence of a digital datum. These physical manifestations of electronic processes are the subject of Valentina Vuksic's art project. In Philipp's submarine station *RemOs1*, we crudely tapped the electromagnetic oscillations in virulent places—power supply, camera activity, onboard computer.[5] And in Hans's case, we complemented recordings of the electronic devices deployed in research with the large-scale infrastructures like the power plant or a drilling tower. In both cases we recorded acoustic signals. The world of acoustic *per se* always refers to time dynamics. For me this is a representation of processuality that is always based on time. One cannot actually pause acoustic signals; they can be perceived only as a stream. I can pause a video, then I have a still image, but when I pause the sound, there is no longer anything to hear. Here an example: The two microphones are located in the basement of the chilling station. They are used to record the signals emitted by the pump stations. These pump stations—as one can see now in the film— convey water to under the roof of the cooling station, where ventilators

---

4   See François-Xavier Standaert, "Introduction to Side-Channel Attacks," in *Secure Integrated Circuits and Systems*, ed. Ingrid M. R. Verbauwhede (New York: Springer US, 2010): pp. 27–42.

5   For audiosamples and the documentation of the setup see the *RemOs1* webpage, https://computersignale.zhdk.ch/en/data/remos1.

rotate to extract heat from the water as it rains back down. The contact microphones record the sound of the pump station. In the captures by the acoustic microphone located directly at the water basin, the signal sounds exactly like strong rain.

Hans Hofmann The sprinkler halls are cooled through evaporation; that's energy efficient in the dry climate.

Hans-Jörg Rheinberger It appears to me that, in essence, you are actually talking about two kinds of materiality: On the one hand, there are the massive infrastructures, including the drilling tower, the chilling station and the power generator. On the other, there is a micro-dimension. You say that we normally think of the digital world as essentially incorporeal, because we do not see the dimension behind it. But every transition from 0 to 1 is ultimately based on a material signal, on some kind of material switch, and not on a Laplacian demon. In the end, the digital world is not incorporeal either. The question is: How are the two materialities related to each other?

Christoph Hoffmann Isn't it about showing the physical processes behind the digital processing? This includes the click from 0 to 1 just as much as the drilling tower. It seems to me that this is conceived less as a chain of translation steps, and more as zooming in ever closer. When they hear the keyword materiality, most people probably think of infrastructures, power plants and drilling towers, or of power and internet cables. They are less likely to think about the physics behind every data processing step. Hannes shows how friction, and ultimately work, take place there as well; computer work. All of these signals that you record make audible what is going on. These sounds are a downright nuisance.

Hans-Jörg Rheinberger On the whole, however, there is only one kind of materiality involved here, namely the materiality of technology. But there is also the materiality of the object of investigation. It disappears here, it is not present at all, no fish, no nothing at all. No genome. Nothing.

Hannes Rickli The fish is there.

Hans-Jörg Rheinberger Where is it? Here, in the aquarium?

Hannes Rickli Precisely. And that is my motive. I realize that with development toward digital science, much more infrastructure comes between the observer and the object of research. My thesis is that the economy of attention is distributed differently. The materiality of the research enterprise is no longer related exclusively to the investigated organism. Thus the focus is shifting to what materially makes up the thread between the animal subject and the observer.

Inside and Outside

Gabriele Gramelsberger The fish is always inside of Hans's laboratory. Its cells are in the refrigerator, the fish is in the aquarium; finally, the fish is in the computer as a datum. The equipment is built up around it. Karin Knorr Cetina wrote a great description of how the relationship between the phenomena changes through the very fact that you bring an object into the laboratory.[6] For Philipp it's precisely the other way around. There the fish is on the outside ...

Hannes Rickli ... which surrounds the *RemOs1* and encapsulates the apparatus. This is why we also integrated the maps into the depiction of the Texas panorama, so that one can show the different places that come into play: There is where the energy comes from, there's where the aquarium is, there's where the data are processed. All in different places, all of them connected with each other over the internet. Philipp's work makes the observer very aware of the 3,000-kilometer long Heligoland-Spitsbergen submarine fiber optic cable. The spatial distribution of the entire research enterprise is an interesting factor. Thus it also concerns time: being at the other location in real time. It is such a tremendous effort, and the fact that it works is anything but trivial.

---

6   Karin Knorr Cetina, *Epistemic Cultures: How the Sciences Make Knowledge* (Cambridge, MA: Harvard University Press, 1999).

Hans-Jörg Rheinberger But what is the difference between this and our satellite system, via which billions of telephone calls are conducted each day, each of which arrives at the right place? Where do these large-scale technological systems, which are even global, differ from a research system? Is there no longer any difference at all?

Hannes Rickli What concerns me: Systems like the internet cannot be individually controlled. Software can still be adapted for a smaller circle, but the internet, the availability of electric power—those are dependencies that have to be negotiated on the societal level.

Christoph Hoffmann I believe we are talking about two different things now. Even around 1800 there were infrastructures. Of course, there were different means of communications and transport, other sources of energy, but research's recourse to some kind of support is nothing new. For the fish facility, a power plant was needed in the background in 1960 as well. But Hannes's interest goes beyond this to the sensualization of certain elements of this infrastructure. It is one thing to show how much infrastructure is necessary for a research project, and another to make immaterial processes sensually tangible.

Hannes Rickli The latter was originally my point. Yet it is an explorative undertaking. I don't just go there and say: Now I'm studying infrastructure. The infrastructures emerged only with Philipp's precarious electricity circumstances. So the infrastructure *per se*—that is, the submarine cable or the drilling tower—emerged empirically as a topic: as something that had to be dealt with.

The Sensual Presence of Data

Hans-Jörg Rheinberger Or from playing with sensors. I would not underestimate how fun it can be to apply devices for detecting noise, to tap one's potential to interfere by implanting little bugs everywhere. That's what you've done. It lends a kind of manifestness to data generation. One hears this apparatus just crackling and creaking and then ponders: How can that be performing accurate work? It acquires a different sensual presence. In the large structures something else is going on. For me, the scientific research process as such, that is, the research question, disappears there, just like the fish that disappeared. That bothers me. Maybe this is even intentional. Then that would be the special trick of the project.

Christoph Hoffmann   You believe that infrastructure cannot be talked about without talking about the objects?

Hans-Jörg Rheinberger   No, it certainly can. But the special challenge here would be to talk about them and, at the same time, to keep the research object and the exploratory research environment present.

Christoph Hoffmann   According to this there are no fixed infrastructures, for infrastructures are also patched together. They recombine over and again with reference to the problem that is currently being worked on.

Hans-Jörg Rheinberger   There are, of course, dimensional shifts. When you make your way to the level of oil towers and large-scale power supply in order to apply your bugs, you are miles away from Hofmann's lab and its neighbors. You can no longer tell the difference between what Hofmann is doing and what Mr. X is up to in the lab next door. The research problems are then no longer tangible, they disappear.

Christoph Hoffmann   But Hannes only goes all the way back to the drilling tower because in the context of Hans's research, nothing happens without the supercomputer, and thus without large-scale energy consumption. In a different constellation this might be completely different. In some projects a fast data connection is a necessity; in others it may not matter at all. So if infrastructure is thought of as a function of the research object, this object becomes very strong. Then the object determines what forms its infrastructure.

Hans-Jörg Rheinberger   But the one supercomputer can work on, let's say, 100 or even 1,000 extremely different research problems. So when you focus on the supercomputer, you cannot tell all of these apart. This is entirely different for the underwater camera.

Infrastructure and Research

Philipp Fischer   One point that seems extremely interesting to me: Does the infrastructure maintain the research, or does the research maintain the infrastructure? Of course, right away I have to think about our research in the polar regions, where it is not entirely clear which supports what. Of course, we build up the infrastructure in order to conduct research. Without infrastructure, we would not be able to survive five days up there. To that extent, the infrastructure is essential. When we were talking about the various perimeters around the nucleus of research, I couldn't help thinking of our cafeteria. It is an integral component of our research up there, it is the center of the whole research life. But the real point I would like to emphasize is, that there are indeed specific phases in which only the availability of a certain infrastructure makes it possible for us to conduct our research. On the other hand, when the infrastructure exists, we are asked to think about research projects for which this infrastructure is suitable in order to justify the operation of the infrastructure. This is a two-sided process, and sometimes it is not so clear what drives what. Building the Neumayer Station, our base in the Antarctic, was the result of political decision-making. We want to belong to the consultative parties that have voting rights in the Antarctic Treaty. One of the conditions for belonging to the club of these consultative parties is that we have a research station operating year-round, which costs us vast amounts of money. Now, since the infrastructure is available, we are asked to submit further projects that are linked to the research station, so that we can better justify it in terms of research. As funny as it sounds, we are a globally renowned research institute, but 25 years ago we were founded primarily as a logistical project with two large-scale infrastructures: the research vessel *Polarstern* and the Neumayer Station. There were 50 logistics specialists and four researchers. Today we have 1,000 people, and even today, 100 of them are logistics specialists.

Hannes Rickli Perhaps you could also say something more about the disappearance of the fish in the infrastructure. Is this a question for you, or is the fish re-embodied in the data? Am I incorrect with the thesis that digitization brings about an increased distance?

Hans Hofmann From our point of view the fish does not disappear, but it is transformed. First it is transformed when it comes from the aquarium in the form of a brain sample or as RNA, or whatever goes into our freezer. And it is transformed anew when the transcriptome of the sample is sequenced, and analyzed again with the supercomputer. For us, I think, the fish—for that is, after all, where our scientific questions come from—is always still quite present. We probably think less about some of the other aspects that are part of Hannes's panorama; for instance, how the cooling functions or where this drilling tower is actually located, etc. Scientists do not have to know this, and it does not necessarily interest them either. To that extent you allow them to see their work in a larger context with which many other projects, many other scientists, can identify. For me the animal subject is not lost, for the simple reason that we pose the questions starting from the animal. I probably have a different relationship to this than those of you looking at it from the outside.

Hans-Jörg Rheinberger Hans says that, in the framework of the analyses he performs, his fish ultimately are present only as RNA samples or as the digital representation of its sequence. This extreme transformation is epistemically interesting. I find this point lacking in Hannes's installation. I would like to be able to grasp this transformation; it should be present in this connection, which the panorama attempts to make audible and visible. From an integral point of view, both poles must be included when talking about research. On the one side, there is the research technology, the framework. On the other side, the frame, as the name says, is the frame for something. And this something is the research object.

Hannes Rickli Perhaps that is why you see this relationship more intact in the capsule under water: because there the images created by Philipp are at the same time present in their raw form. In contrast, the gene sequence that is the objective of Hans's research is no longer present, not even as a text.

Christoph Hoffmann Although Valentina would say that the text could be reconstructed from the registered computing operations.

Hannes Rickli Yes, theoretically.

Hans-Jörg Rheinberger That would be super. You show a video sequence from the TACC and out of the soundscape come nucleotides.

Stability/Instability of Infrastructures

Gabriele Gramelsberger What I found so beautiful about Hannes's panorama is that one sees how the fish is wrapped in something else, and this something else, for its part, is wrapped in infrastructure, and this infrastructure is wrapped in an even larger infrastructure. Right, the wrapping. A tremendous number of calculations are running on the computer, but when the drilling tower stands still and the energy supply collapses, the cooling stops working. Then the computer breaks down, and the data may be lost, meaning that the research is lost: it dissolves into oblivion. This dependency becomes just as clear through Philipp's power outlet. Normally one does not think about it all here. This stupid seawater is so corrosive, it doesn't do what we want, while mankind generally has everything else pretty well under control.

Hannes Rickli It does a lot more. As part of bioactivity it produces algae on the viewing windows. It attacks the windows, it attacks the housing.

Christoph Hoffmann That's a reminder of the vulnerability of infrastructures. In Philipp's work one always has the impression that the infrastructure is constantly in danger. There is always something broken. While in Hans's work the interconnections appear to be more stable?

Hans Hofmann Now and then there are problems, but the people in the computing center communicate very well, etc. Nothing that worries me.

Christoph Hoffmann The worst would be if there were a power outage.

Hans Hofmann Yes, but there is never a power outage on the UT campus.

Hans-Jörg Rheinberger Because there are two backup systems?

Hans Hofmann No, because we have our own gas power plant. We have a pipeline coming in from West Texas; we are completely independent from the rest of the world. Two years ago we did have a power outage, however, which should not have happened. Construction workers destroyed some main cable. Apparently, that was the first time since 1981 that there had been a power outage on the UT campus, and it lasted about a half hour. The freezers can break down, of course—that has happened. Our building has emergency freezers, into which everything has to be moved in such cases. But as long as you don't open the freezer, it stays cold for hours. When you know where you are bringing the contents, you get some dry ice in polystyrene containers. Then everything is transferred relatively efficiently, and I have to buy a new freezer for 12,000 dollars. That has happened to us twice already in the last five years. TACC, I would say, is very stable. They have 140 people, making it one of the largest civilian supercomputers in the world, they have a huge budget. Our sequencing facility is quite stable. The instability comes from the fact that the technology continues to develop ever further and you don't want to be left behind. And the instability may also come from the fact that I would like to have my data within one or two weeks, but under certain circumstances they may take three or four weeks.

Hans-Jörg Rheinberger Getting back once more to the differentiation between data generation and data processing, what we have at the station off Spitsbergen is an extremely fragile situation for data generation. This is different in Hans's case. There the data generation is embedded in all kinds of routines. Occasionally a part may fail, or if a brain sample gets lost, he just takes the next fish; he has enough fish. This situation, this context in which the data are generated, is dramatically different compared to Philipp's situation.

Philipp Fischer   One has to see this in the correct context. In Hans's institute, he has control of 90 percent, maybe even 100 percent of the environmental variables that complicate his work. He can create redundancies, the University of Texas has a power plant of its own providing energy on demand and he can make the lab operation completely independent. That's why it is a stable system. Our situation is, unfortunately, quite different. I cannot influence the glaciers. When an iceberg comes down that is 80 meters long and extends 30 meters into the water, I cannot create any redundancies in order to prevent the destruction of my underwater station. If I set up 25 machines, 25 machines are destroyed. I have control of 50 percent of data collection, but no more. When the environmental constraints become really critical for our research in the field, and we ask ourselves, sometimes almost with tears in our eyes, whether all these challenges are worth it, then my favorite saying is: "Nothing ventured, nothing gained." With this approach we at least relieve some of the pressure we're under.

Hans Hofmann   The most fragile thing we have are the fish. And we have had larger accidents that cost us many animal subjects.

Hans-Jörg Rheinberger   Because they became ill?

Hans Hofmann   Yes, either some disease sweeps through, or something in the life support system goes wrong, despite the built-in redundancies. In the last ten years, since I have been at UT Austin, we've probably had three or four such cases, not all of them so extreme. In the worst case it set us back six months. I would say that this first step is the most fragile. Here we are still closest to nature and have the least control.

Philipp Fischer   For us it's exactly the other way around. When a fish swims by me, I don't know whether it's eaten the next day or not, and that has no influence on the results.

Hans-Jörg Rheinberger The main thing is, it came along.

Hannes Rickli And was captured by the system.

Christoph Hoffmann So if vulnerability does arise, then, in contact with nature? In the case of Philipp's observation station, then, in the contact with saltwater, with the low temperatures.

Gabriele Gramelsberger Or for Hans, in the contact with the fish.

Christoph Hoffmann In drastic terms, the acquisition of the research object is accompanied by saving it from disturbance, decomposition and destruction. The further the object is extracted from its natural environment, the more ...

Hans Hofmann ... stable it is.

Philipp Fischer We have to be careful here. The entire system does not become more stable due to the fact that the object is extracted from its environment; the danger is merely of a different kind. When I look at Hans's approach and my own, my object of research as such is not threatened at all, since it is unlikely that the Kongsfjord in Spitsbergen will collapse in any form and that all organisms there will die. For me the infrastructure is the essence of insecurity. The more Hans extracts his research object from the water and brings it into his infrastructure, the more problems he has with the safety of his object.

Hans Hofmann Even in an experiment that works, you are confronted with all of the variability you have with living organisms; with this unpredictability. My students and postdocs always breathe a sigh of relief when the tissue samples are in the box. From there on they feel that they have much more control over their object. They are no longer driven.

Christoph Hoffmann While it would be no problem for Philipp, it would not be good for the ecosystem if an oil spill were to happen. You would just have a different ecosystem.

Philipp Fischer   That would certainly be quite dramatic, but it would not take long for us to draw up a new research question and continue working.

Hans Hofmann   That is exactly what happened in the Gulf of Mexico after the BP oil spill. I know people who were working there on all kinds of projects. From one day to the next they had a different research project, although they had not changed a thing. Those who were clever exploited this very well and learned a great deal.

Christoph Hoffmann   Earlier we were saying that the distance between the research object and the researchers is greater today, and the space between much more filled with technologies without which the research object would not be available. In the beginning I was still working without a computer, with books, libraries, a typewriter ...

Gabriele Gramelsberger   ... Tipp-Ex ...

Christoph Hoffmann   ... pen and paper. Today my research environment, put bluntly, consists of the internet and a computer that is replaced every five years. But it seems to me that the bond to the infrastructure always remains the same. Technically, it is certainly more complex today: I cannot unscrew the computer, I have no idea how to do that. I have a problem when my internet connection breaks down. But in 1985 it was a crisis for me when the library was closed for Christmas break.

Distancing through Infrastructures

Hans-Jörg Rheinberger   That could have something to do with the fact that there are anthropological constants that limit how much an individual can manage and pay attention to. The infrastructure may become ever more complex, but the more complex it becomes, the more it has to be black-boxed. Your attention, your very possibilities for taking up and processing information, are limited. I am certain that our human attention potential is somehow reflected in these highly complex infrastructures. Otherwise one would face a situation in which one is simply inundated and basically paralyzed. When I look back on my time in the laboratory, it was just the same. In order to be able to concentrate attention on the research question, I had to presume—take for granted—that the ultracentrifuge would do what it was supposed to do. There may have been the occasional accident, but otherwise this piece of technology had to be taken as unproblematic. I think this is basically no different today.

Hannes Rickli Infrastructure is hidden from view. That is its characteristic.

Hans-Jörg Rheinberger But if the infrastructure is modular, you can always open it at certain places. It doesn't work without such modularity.

Hans Hofmann In a certain way Hannes shows this modularity. The power plant can be replaced without replacing the computing center, or the fish facility, or the laboratory, etc.

Hans-Jörg Rheinberger That goes all the way into the details. It's the same thing if you focus on the computer.

Hans Hofmann What's called hot swapping takes place when a certain drive stops working: You can exchange and replace a single hard drive without having to turn off the supercomputer. There's one person who has this single task, who moves through the *Stampede* computer all day replacing these hard drives, because they only have a limited service life.

Gabriele Gramelsberger In Heideggerian terms, Hannes visualizes the data's "readiness to hand." Heidegger differentiates between readiness to hand (*Zuhandenheit*) and presence-at-hand (*Vorhandenheit*).[7] That which is present is what you no longer perceive because it works. At the moment it no longer works, it becomes present; you now perceive it. Hannes makes this presence-at-hand accessible.

---

7   Martin Heidegger, *The Question Concerning Technology and other Essays*, trans. William Lovitt (New York: Garland Publishing, 1977).

Hans-Jörg Rheinberger It seems to me that science, as a whole, is based on an estrangement of this kind, but in order to realize it, paradoxically, one has to forget it.

Hannes Rickli I am interested in this from the perspective of aesthetics, of perception. This concerns not only processes of research, for much more general questions are involved. What's interesting is that one can study the differentiation. Completely different dimensions and things count for Hans than for Philipp, for whom electricity or bioactivity, for instance, become palpable very directly, quite essentially.

Hans Hofmann And where does aesthetics come into this?

Hannes Rickli I understand aesthetics as work on perception. I am aware whether something can be sensually perceived, and I work at the gaps, on those places where things elude perception.

Gabriele Gramelsberger In media theory there is a great deal of discussion about the "technological unconscious."[8] What this means is that technologies make the unconscious visible on the one hand, but on the other, become themselves unconscious through the embedded infrastructures.

Hans-Jörg Rheinberger I think, Hannes, perception alone is still not quite enough. I believe that what also essentially belongs to this, and what constitutes aesthetic effects, is a certain degree of surprise. Without surprise—a kind of wonder—there is, I believe, no aesthetics at all. But without a certain degree of surprise there is also no epistemology, that is, knowledge. This is where the production of epistemic effects meets the production of artistic effects.

Hans Hofmann Perhaps most accessible intuitively is Hannes's installation with the Kramer sphere.[9] Visitors to the installation went from one

---

8   Nigel Thrift, "Remembering the Technological Unconscious by Foregrounding Knowledges of Position," *Environment and Planning D: Society and Space* 22, no. 1 (2004): pp. 175–190.

9   Hannes Rickli, *Spurenkugel: Ein Schreibspiel* (Baden: Lars Müller, 1996).

surprise to the next: The thing revolved, one attempted to understand what was happening. Tracks were generated, a computer evaluated and printed the data.

Hans-Jörg Rheinberger The generation of uncertainties—suspense, insecurity—manifests itself in all of these aspects.

Hans Hofmann At the time I enjoyed the process of discovery. I go through there and influence the entire process by activating the light barriers. That was not evident for the user from the start. Something happens, in a certain way the process of research is reconstructed when there is actually a cricket or some other kind of creature sitting on the sphere. Yet, at the same time the process is also reversed. The same thing happens in the panorama, only the experimental system is much more complex, has many more dimensions and extends much further spatially and temporally.

Hannes Rickli When one compares these with each other (and we could argue about whether or not that's productive), the Kramer sphere was actually an illustration of the research process. What happens in the Austin panorama is something different, which is not symbolic, as it aims directly at the presence of the phenomena—that is, the technical facilities—and thus at sensual experience. I noticed that the infrastructures bear an aesthetic potential even without being symbolically transferred and validated. The actual artistic work here consists in expressing their own materiality and energetic power.

Hans-Jörg Rheinberger There was much more to discover in the sphere than you had intended. But that will happen to you again with the panorama. Works of art, like objects of research, point beyond themselves.

In silico

The Developmental Hourglass

Hans Hofmann We have now on several occasions touched on the importance of data available from public repositories for pursuing fundamental
biological questions with data-driven *in silico* analyses. I want to illustrate
this approach with a current project in my laboratory, in collaboration with
Dr. Becca Young at UT Austin, where we analyze large, publicly available
datasets with new methods, thus permitting new insights into traditional
biological questions. The specific research, which is illustrated by the
poster you see here (supplement 1), aims at an old biological problem that
remains unsolved to this day. I chose this example because it has a historical context that may be of interest to you. In the mid-nineteenth century, Karl Ernst von Baer investigated embryos of many different species
under the microscope and determined that they looked very different at
the beginning of embryonal development and later as well. But between
these phases there was a period during which they were quite similar.[1]

Hans-Jörg Rheinberger With regard to mammals?

Hans Hofmann Originally with regard to vertebrates, but the same observation was made later for worms and arthropods as well. So this is not restricted to vertebrates. At some stage this phenomenon was dubbed the
developmental hourglass. The question was, what led to this reduction
of variability, and which mechanisms are at work? In the 1920s this was
affirmed once more, and in the last decade a few studies have been published in which expression profiles were examined genome-wide, known

---

1   Karl Ernst von Baer, *Über Entwickelungsgeschichte der Thiere: Beobachtung und
     Reflexion*, 2 vols. (Königsberg, 1828/1837).

as transcriptomics.[2] Here it became apparent that the similarity of the gene expression profiles—shown here as the correlation coefficient—is significantly greater during this period than in the earlier or later embryonic stages. This period was designated by Klaus Sander as the phylotypic stage (or phylotypic period).[3] It is typical for a phylum, for instance, the vertebrates. Judging from these papers, this appears to be the molecular basis for the phenomenon. What interests us is: What happens here, and why does it happen this way? Rudy Raff, who comes from the evo-devo context—a research area focused on how the control of individual development took shape through evolution—introduced a hypothesis on this in the 1990s.[4] According to Raff, gene expression networks, that is, the functional networks, are relatively isolated from each other early and late in embryonic development. In the phylotypic stage, by contrast, in which the embryos look similar, they are highly integrated. The basic idea behind this hypothesis is that due to pleiotropy—that is, when one gene

2   Alex T. Kalinka, Karolina M. Varga, Dave T. Gerrard, Stephan Preibisch, David L. Corcoran, Julia Jarrells, Uwe Ohler, Casey M. Bergman, and Pavel Tomancak, "Gene Expression Divergence Recapitulates the Developmental Hourglass Model," *Nature* 468 (2010): pp. 811–814; Tomislav Domazet-Lošo, and Diethard Tautz, "A Phylogenetically Based Transcriptome Age Index Mirrors Ontogenetic Divergence Patterns," *Nature* 468 (2010): pp. 815–818; Naoki Irie and Shigeru Kuratani, "Comparative Transcriptome Analysis Reveals Vertebrate Phylotypic Period During Organogenesis," *Nature Communications* 2 (2011): article 248; Itai Yanai, Leonid Peshkin, Paul Jorgensen, and Marc W. Kirschner, "Mapping Gene Expression in Two Xenopus Species: Evolutionary Constraints and Developmental Flexibility," *Developmental Cell* 20 (2011): pp. 483–496; Alex T. Kalinka and Pavel Tomancak, "The Evolution of Early Animal Embryos: Conservation or Divergence?," *Trends in Ecology & Evolution* 27 (2012): pp. 385–393; Marcel Quint, Hajk-Georg Drost, Alexander Gabel, Kristian Karsten Ullrich, Markus Bön, and Ivo Grosse, "A Transcriptomic Hourglass in Plant Embryogenesis," *Nature* 490 (2012): pp. 98–101; Naoki Irie and Shigeru Kuratani, "The Developmental Hourglass Model: A Predictor of the Basic Body Plan?," *Development* 141 (2014): pp. 4649–4655.
3   Klaus Sander, "The Evolution of Patterning Mechanisms: Gleanings from Insect Embryogenesis and Spermatogenesis," in Brian C. Goodwin, Nigel Holder, and Christopher Craig Wylie, eds., *Development and Evolution* (Cambridge: Cambridge University Press, 1983): pp. 137–159.
4   Rudolf A. Raff, "Developmental Mechanisms in the Evolution of Animal Form: Origins and Evolvability of Body Plans," in Stefan Bengtson, ed., *Early Life on Earth* (New York: Columbia University Press, 1994): pp. 489–500; Rudolf A. Raff, *The Shape of Life: Genes, Development, and the Evolution of Animal Form* (Chicago: University of Chicago Press, 1996). Duboule had a similar idea: Denis Duboule, "Temporal Colinearity and the Phylotypic Progression: A Basis for the Stability of a Vertebrate Bauplan and the Evolution of Morphologies Through Heterochrony," *Development*, Supplement (1994): pp. 135–142.

influences many different mechanisms or controls many different processes—it is not so simple to change a gene's activity through evolution. This consequence of this constraint is that the embryos look relatively similar—as far as both gene expression and morphology are concerned.

Hans-Jörg Rheinberger On what basis did Rudy Raff develop this hypothesis at the time? He certainly did not have the technologies and the data that we have at our disposal today.

Hans Hofmann In the early 1990s he simply dug up the old literature and said: We have a problem, we have to explain this—and here is a hypothesis. He had no evidence or data basis of any kind. The matter was then forgotten again, until the technology was finally ready about a decade ago. But until now nobody has looked at these networks and tested the hypothesis in practice. And that's precisely what we want to do.

Christoph Hoffmann What do you do in order to test this hypothesis?

Hans Hofmann Practically every individual gene and its activity over the period of development have to be measured. So transcriptomics, applied to different species. With our partners at Michigan State University we are also working on *in silico* evolution. They developed Markov network brains, which are, in practice, agents. Each cell represents one brain. Imagine this as a tissue that develops and forms certain patterns; these would be the phenotypes. Then evolution can be run *in silico*, in the computer, to see whether or not the development of these artificial-life organisms in the computer resembles this hourglass phenomenon: whether it is thus generally a characteristic of development processes. This is a complementary approach, quite new in developmental biology and evo-devo. These areas of research are generally lacking in theory and have relatively few quantitative approaches, aside from the classical reaction-diffusion models (for instance, the Gierer-Meinhardt model). For the transcriptomes you have the problem that you have to identify the orthologous genes—that is, the corresponding genes—from all of the different species under study. We do this with comparative analyses, but there are also other possibilities. In the analysis I will show you now, we looked at the data over the development period for 2,500 genes—these are actually orthologous gene groups, but let's just call them genes for now—for various species; at the moment, only for vertebrates. They make up just over ten percent of the genome, which isn't bad. We have the largest dataset that has ever been analyzed genetically; all of them data that were downloaded from public databases. These are both microarray and RNAseq data, which we then integrated in various ways.

Christoph Hoffmann You leave out the integration step because it would take too long?

Hans Hofmann I leave out the data integration because it gets quite detailed. Now you can see that there is a change in the correlation which measures the similarity of the gene expression pattern from the early to the late embryonic stages (supplement 1, center column, above). But in the pharyngula stage, which is generally viewed as a phylotypic stage, all are relatively similar to each other. The r-values indicate that they do not differ from each other statistically, while here the similarity decreases quite a bit. With the larger dataset we can thus see that the hourglass cannot be readily reconstructed on the level of expression. One can also combine and summarize the data differently into Early, Middle and Late stages. Here, too, one sees that Early and Middle have roughly similar correlation coefficients. However, there is much more variation in the Early stage than in the Middle. This could be an indication, but compared with what others have published with smaller datasets, it does not seem to be a very robust signal. Then we decided to first reduce the complexity of the dataset—for they contain hundreds of millions of data points. We do that here with a Principal Component Analysis (PCA), which identifies the components that explain fractions of the total variance in the data (supplement 1, center column, middle). The main component, which contains the largest share of total variance in the dataset, is PC1 (principal component 1). Eighty-three percent of the variation in the data is attributed to PC1. This is different for the various species: fish and mouse, chicken, and here on this side the frog species *Xenopus laevis* and *Xenopus tropicalis*. Principal components 1 and 2 differentiate the various species, as it were, and thus explain the differences between the species. If you then look at principal component 3, you see how the individual development stages from 1 to 15 line up. PC3 represents only 3.1 percent of the variation, but explains the variance over time. This component thus represents the data according to the different development stages. Principal components 1 and 2, for their part, explain the differences between species, especially between the mammals, and between the fish and the frogs on the one hand, and the chicken on the other. Most interesting for our purposes, however, is principal component 3. It accounts for just three percent of the variation, but precisely these three percent explain the development stages from 1 to 15.

Hans-Jörg Rheinberger And what does a component mean?

Hans Hofmann That's a component of the variance in a dataset. The objective of principal component analysis is to approximate the multitude of statistical variables in a complex dataset through a lower number of potentially significant linear combinations—known as the principal components. Each variable (for instance, the expression level of the investigated gene) "loads" on one of these main components.

Hans-Jörg Rheinberger Are components gene clusters?

Hans Hofmann Yes, in a sense they are: Every gene loads onto a certain principal component, and thus contributes to a certain principal component. There are a couple of hundred genes that load primarily to PC3; that means they contribute to this temporal structure in the gene expression profile. And we think that these genes may show the hourglass behavior.

Hans-Jörg Rheinberger Can you identify them?

Hans Hofmann Yes, there are a couple of hundred. I'll just show you here how such a gigantic dataset can be broken down in practice and observed from various perspectives. There is an unbelievable amount of information here. A value of 3.1 percent of the variation doesn't sound like much, but this could potentially be the genes we're interested in. The important axis is the Y-axis (supplement 1, center column, middle).

Christoph Hoffmann And that is a time axis?

Hans Hofmann No, although it looks like one; that is the principal component, i.e. one component (of many) of the variation in the dataset. Unexpectedly for us, the Y-axis here reflects the time axis: from 1, early embryo development, to 15, the late embryonic phase—which indicates that the genes that load on this component are the ones that explain the temporal structure and change during embryonic development.

Christoph Hoffmann That means that the temporal sequence emerges from the analysis; isn't that what is assumed?

Hans Hofmann Exactly. That's why it was so surprising. We did not expect to see that. These multivariate analyses are good for recognizing in the data patterns what would otherwise not be seen at all, or would not be expected. In this case we then attempted to relate the various embryonic stages of the different species to each other. For the chicken, for example, everything from 1 to 15 is compressed, whereas it is more stretched out for the other species. The question is: How can they be related to each other, what are actually the equivalent embryonic stages? This is a relatively difficult problem. First of all—this is all a work in progress, this is a poster from a recent conference—we looked at the classical literature to see when the various development stages described here appear during development in relation to this somite pair. For every embryo is divided into segments called somites. This is true not only for arthropods, but also for vertebrates. The question is how the development stages of one species can be matched to the stages of other species. We now take this kind of information and use it to align the embryonic stages of each species with the corresponding stages of the other species. In practice, morphological information is linked with the gene expression. When you now break down principal component 3, which reflects the embryonic period in development, across the stages of development, you see a more or less linear relation here. There is still a minor error, but overall the relationship seems to be fairly strong between gene expression, which varies over time, and the morphologically defined embryonic stage. This is the current status. We have identified the genes that load on this principal component here, and can now perform additional analyses. The trick with these network analyses is that we need quite large samples, at least 20 or so, to make them statistically robust. We have achieved this for some of the embryonic stages. Everything we've used comes from public databases, and for this it was necessary to integrate data from different technology platforms.

Quality of a Dataset

Christoph Hoffmann You take the datasets from public repositories, and this brings us back to the question of trust: How reliable are the downloaded data? Is this a question at all at the moment, or do you discuss such questions *ex post*?

Hans Hofmann No, this is a question from the very beginning, for if you have bad data, you do not want to waste time with them. This starts with various quality-control analyses—that means we look to see which data comply with even the minimum requirements. None of the others are included in the analysis at all. There are a few datasets we excluded.

Christoph Hoffmann And the community has common standards concerning the minimum requirements?

Hans Hofmann Yes, certainly on this level. That does not mean that everyone adheres to them; obviously, crappy data are also published.

Hans-Jörg Rheinberger But how do you decide about the quality of a dataset? It does not carry a quality label when it becomes publicly available. I could imagine that one simply uses them, and if it turns out that there is something odd about them, they are thrown out again.

Hans Hofmann That is our second step. It may be that a dataset passes the first quality control and it then turns out to contain a systematic error, so that we have to exclude the dataset in question. The first quality control checks only whether they are data that were produced in the normal way by the corresponding technology, and thus have the usual distribution. You can check this before the analysis. The quality control always consists of multiple steps. It could be, for instance, that the data for development stage 5 are somewhere down here (supplement 1, center column, middle). We have seen something like this. If you look at the data as a whole, you know that something probably went wrong. You would probably take another look at them first, to see whether you might have made some kind of systematic error yourself. If that is not the case, perhaps the researchers who put the data into the repository made an error during uploading. If that is not the case, something may have gone wrong with the measurement, etc. Possibly the error can be found and eliminated. In any case one has to be quite meticulous with such operations. But back again to *in silico* evolution. Anyone who lets the development of these artificial organisms run as programmed sees that they can produce very different patterns. Development can take place, that's the first thing nobody has ever done. If you then apply something like natural selection and say that you would ultimately like to have a certain pattern as your phenotype, and then run that program on the computer 10,000 times, it also becomes clear that such patterns sometimes appear relatively quickly, while sometimes it takes much longer. In the development of animals and plants one would call this heterochrony. What's interesting is that this heterochrony also occurs in these model organisms. Thus this is apparently an emergent property, meaning a property of the development of organisms that takes effect quasi automatically.

Agent-Based Models

Hans-Jörg Rheinberger   So is this work essentially based on game theory? Or are other theories involved?

Hans Hofmann   Yes, they tend to be agent-based models. In practice, every agent works as a simplified brain; every agent is a cell in the brain.

Hans-Jörg Rheinberger Does it, in principle, work the way Manfred Eigen and Ruthild Winkler described it in their book *The Law of the Game*?[5]

Hans Hofmann Yes, but here it is not a game theory approach. Game theory would not be suitable for this, because it is not agent-based; rather, it provides numerical solutions for an entire population.

Hans-Jörg Rheinberger But you roll the dice …

Hans Hofmann You generate variations, and you can do that by rolling the dice; here we call it mutations.

Hans-Jörg Rheinberger And what is operational when you say agent-based?

Hans Hofmann Every individual cell here is an agent (supplement 1, right column, above). There are very simple rules. For example: If the neighbor is so, than so am I. Only there are two, or three or four rules. They are the same as for a biological process. You generate mutations, the mutations generate different phenotypes, and these are affected by the selection acting in the computer. You can define it. On the computer you can play God, if you want to.

Gabriele Gramelsberger But are mutations neighborhood-based? Does that mean you can describe mutations through neighborhood rules?

Hans Hofmann You could, but that's not how it works in our case. Something changes in a cell coincidentally, and that has consequences for the neighbors.

Christoph Hoffmann But you use this approach exclusively in order to investigate the emergence of certain patterns.

Hans Hofmann Yes, first of all we want to see whether such patterns

---

5   Manfred Eigen and Ruthild Winkler, *The Law of the Game: How Principles of Nature Govern Chance*, trans. Robert Kimber (New York: Alfred A. Knopf, 1981).

develop, and second, how they develop and whether or not something like an hourglass emerges in the long term.

Christoph Hoffmann   And if the hourglass were to emerge, would that perhaps be a condition of development?

Hans Hofmann   That would be one possible interpretation. If there is supposed to be such an hourglass, we could, moreover, mathematically determine the correlations between all of these genes, in order to see whether Raff's hypothesis of networks that are either more or less integrated would be applicable here as well. But we're not that far yet. The first thing we discovered was that when patterns are selected, these patterns actually do form reliably through evolution. Second, we discovered that there is heterochrony. And third, that quite similar end products may develop again when these experiments are repeated. You see that here, when you compare right and left (supplement 1, right column, middle). This observation goes back to Stephen Jay Gould, "replaying the tape of life."[6] This also happens with the Markov network brains in the artificial life project. This is the current situation. We have not come far enough to see whether there is an hourglass and what that looks like. In summary: First we downloaded datasets from public repositories and massaged them, meaning, we changed them so that they were comparable to each other.

Data Massage

Gabriele Gramelsberger   You "massaged" the datasets?

Hans-Jörg Rheinberger   Ah, now we're coming to data massage. Yesterday we were dancing with the data.

Hans Hofmann   We massaged the data so that they could be analyzed to-

---

6   Stephen Jay Gould, *Wonderful Life: The Burgess Shale and the Nature of History* (New York: W.W. Norton & Company, 1989).

gether. Then we checked whether we were able to reproduce what others had found before us. We can do this only in part. Perhaps because we have a very complex dataset and only part of the dataset, a part of the genes, is at all relevant for the pattern we are seeking. Therefore the next step was to perform the PCA and identify a set of genes that actually reflect the embryonic development. Now we have to proceed from this point. We match the embryonic development on the level of gene expression to the morphological development. This is still a work in progress and to be optimized further. But here one sees that there is already a quite linear relation between the gene expression and morphological development (supplement 1, center column, below). We completed all of this with the Markov network brains, and were thus able to show that they, too, can develop. This has never been done before. Generally these models are used for other purposes, not for questions in developmental biology. That something like heterochrony could result as an emergent property was not our question at all. Nor did we expect to repeatedly find evolutionary processes that lead to similar results. Overall it is a mixed approach: You have data parasitism, then the fishing expedition approach in the form of exploration; plus, in the same study, you have hypothesis testing complemented by the *in silico* modelling approach. This is representative of much of what is happening today, and will happen tomorrow, in my research and in the whole sphere of big data genome-relevant research. This includes a systems biology approach, developmental biology, and comparative evolutionary biology.

Hans-Jörg Rheinberger   A bit of everything, so a single perspective is not enough. You said at the beginning that around 2,500 genes are represented in the dataset?

Hans Hofmann   For which we have data from each individual dataset.

Hans-Jörg Rheinberger   How many of these are attributed to PC3?

Hans Hofmann   Slightly more than 200, I believe. Just under ten percent, 210 or so. If one looks how many genes generally engage in development—as transcription factors or in other functions, not as mere fellow travelers, but taking on a causal role, then the estimates in other studies range from ten to fifteen percent. These may be the ones sought, or maybe not. But the order of magnitude is more or less right.

Hans-Jörg Rheinberger That would fit. You say that the 2,500 are around ten percent of the entire set of genes. And this number reflects the set of genes that are relevant for development?

Hans Hofmann We have good reason to assume that these 2,500 genes are representative. No bias is implicit. There are all kinds of possible reasons why we cannot find orthologs for the other genes. It appears to be random. A number of 2,500 is quite good for a comparative study spanning 450 million years of evolution. In another study, we look at the evolution of mating systems, that is, social systems; there we have about 2,100.[7] In the paper by Irie and Kuratani, in which the expression similarity for the pharyngula stage was shown, the data for only 300 or 400 genes are considered.[8]

Christoph Hoffmann Do you trace your results back to the additional amount of data considered?

Hans Hofmann I'm not sure how it can be explained, I don't know.

Christoph Hoffmann So you are not giving up on the initial hypothesis of the hourglass effect?

Hans Hofmann No, I think that these 200 genes that load to PC3—this is my hypothesis now—reproduce this pattern here. This is where we stand now. If you compare the two figures here (supplement 1, center column,

7   Rebecca L. Young, Michael H. Ferkin, Nina F. Ockendon-Powell, Veronica N. Orr, Steven M. Phelps, Ákos Pogány, Corinne L. Richards-Zawacki, Kyle Summers, Tamás Székely, Brian C. Trainor, Araxi O. Urrutia, Gergely Zachar, Lauren A. O'Connell, and Hans A. Hofmann, "Conserved Transcriptomic Profiles Underpin Monogamy Across Vertebrates," *PNAS* 116, no. 4 (2019): pp. 1331–1336.
8   Irie and Kuratani, "Comparative Transcriptome Analysis Reveals Vertebrate Phylotypic Period during Organogenesis."

above), you see, however, that what has been published is not so easy to reproduce. We looked for a simple answer, and found various complex answers. Now we believe that our datasets are large enough for such network analyses. But then you first have to prove statistically that these analyses are robust enough that you can also believe the results.

Gabriele Gramelsberger   But your studies do not reflect this network topology, do they?

Hans Hofmann   We are not there yet. We have only shown that *agent-based models* can actually show development phenomena and reflect some of the phenomena that are also seen in embryonic development. The next step is to see whether this hourglass appears there as well. Everything that has been shown so far serves as the basis for being able to test the hypothesis at all. And that's why I say: It's a work in progress.

Why *in silico?*

Christoph Hoffmann   One question: Why do you do this *in silico*?

Hans Hofmann   One reason is that we have only one hypothesis, a H1, that there is pleiotropy and that this network integration is either less strong or stronger. This is not satisfactory. I don't like any experiments here. Generally I prefer multiple hypotheses, and then either have support with strong inference for these or not. If I also have organisms *in silico* that are being affected by evolution, and over which I have much more control than over animals and plants—especially since I can have the process repeated on the computer—and I then find similar conditions, that further reinforces my hypothesis.

Christoph Hoffmann   Or a new hypothesis emerges.

Hans Hofmann   That's the other possibility. This approach will probably supply us with new hypotheses which we can then test with biological data. To date we are able to reproduce part of what is in the literature, but not all of it. We can explore the data and reduce complexity so that we may perhaps be able to understand why we can reproduce the literature and why we cannot. Through comparison with the morphological development we can improve the data such that they are more representative biologically. And we can develop an independent system in which evolution also takes place, in order to see whether similar phenomena arise. This is the combined, integrative approach.

Gabriele Gramelsberger This hourglass doesn't really make sense to me. Where does the variation in the embryonic development on the left edge come from (supplement 1, left column, above)?

Hans Hofmann That's a good question. There are various possible reasons for this. There you see early embryonic stages, the eight-cell stage, the sixteen-cell stage ...

Gabriele Gramelsberger But in terms of morphology, they are identical?

Hans Hofmann No they aren't. They look quite different from each other. In general, embryos are very different at the beginning. There are two main reasons for this: They may have been affected by natural selection, because the embryos are in different environments, and selection ensured that they were optimally adapted and thus look and work differently. Or it could be, precisely the opposite, that during early embryonic development the embryos can vary more or less without being affected by selection. Along both paths you obtain variability. This variability can be explained in part by selection, and in part by independent evolution. For example, the fish here and these mouse embryos had their last common ancestor 450 million years ago. Since then they have experienced a different evolutionary history. On the basis of the different trajectories alone, you would expect them to be different, even if no selection had influenced them. But selection did influence them, of course, which explains part of the variation in morphology. There are statistical methods to analyze how much of this variation can be explained by selection, how much by random processes, and how much by phylogenetic constraint.

This designates a component of the phenotype of a lineage that prevents or restricts any other possible (or even expected) evolutionary adaptation. It is also important that there are different reasons why it varies to this extent here, and to that extent there. But what we want to know is why there is much less variation during the phylotypic stage in terms of morphology, and potentially also in terms of molecular biology. If we presume that the early embryonic stages are similar for all species, this is the fault of Ernst Haeckel. Haeckel misinterpreted von Baer's embryological observations when he proposed his phylogenetic law.[9] But this has not played a role in science for decades.

Hans-Jörg Rheinberger   The general image is very different. Haeckel's idea is: Ontogenesis repeats Phylogenesis.[10] This means that the embryos are quite similar at the beginning and then gradually develop in opposite directions. So, not an hourglass, but rather a spreading from a shared starting point.

Hans Hofmann   That's why I said that Haeckel misinterpreted von Baer. Von Baer could no longer counter him effectively. In the late 1860s he was an old man. When Haeckel's phylogenetic basic rule was discredited, von Baer was thrown out along with it. Only in the 1960s did Friedrich Seidel confirm von Baer's observations, which prompted Klaus Sander in 1983 to introduce the concept of the phylotypic stage.[11] In the early 1980s, Rudy Raff, too, said: Perhaps we should read von Baer again; he wrote down a problem for us that we still have yet to solve. Raff is a brilliant guy. Knows an unbelievable amount, about history as well. I have not discussed this project with him, it is all still relatively new. He published

---

9   Søren Løvtrup, "On von Baerian and Haeckelian Recapitulation," *Systematic Zoology* 27 (1978): pp. 348–352.

10   Ernst Haeckel, *Allgemeine Entwickelungsgeschichte der Organismen: Kritische Grundzüge der mechanischen Wissenschaft von den entstehenden Formen der Organismen begründet durch die Descendenz-Theorie*, Generelle Morphologie der Organismen, vol. 2 (Berlin, 1866): p. 300.

11   Friedrich Seidel, "Körpergrundgestalt und Keimstruktur. Eine Erörterung über die Grundlagen der vergleichenden und experimentellen Embryologie und deren Gültigkeit bei phylogenetischen Überlegungen," *Zoologischer Anzeiger* 164 (1960): pp. 245–305; Sander, "The Evolution of Patterning Mechanisms."

his essay in 1994, but, again, not much happened until the study by Irie and Kuratani appeared in 2011.[12] That was the first one, and now there are three or four such publications with somewhat different approaches, all of which were published in high-profile journals. Irie and Kuratani looked only at vertebrates. In the meantime, similar patterns have been found for plathelminthes and for arthropods. But all of this is exploratory and descriptive; nobody has tested any kind of hypotheses yet.

Rearranging

Christoph Hoffmann Back to the poster again. I am interested in the step from the center diagram to the lower one (supplement 1, center column). What happens in the data analysis there? In the diagram in the middle, the data are read according to a certain question and then rearranged in a graph.

Hans Hofmann They are rearranged, but their complexity is also reduced. From millions of data points, we arrive at, I don't know, a few dozen.

Christoph Hoffmann Would you say that the next step, from the center downward, is accompanied by a further rearrangement? Or is it, rather, adding additional information to the graph? I am trying to understand to what extent the data are increasingly linked with meanings from one step to the next.

Hans Hofmann The reduced data are now compared with the other dataset. Here another information increase takes place.

Christoph Hoffmann Could one say that this step is also accompanied by a further theorization?

---

12 Raff, "Developmental Mechanisms in the Evolution of Animal Form"; Irie and Kuratani, "Comparative Transcriptome Analysis Reveals Vertebrate Phylotypic Period during Organogenesis."

Hans Hofmann Rather than theory, we would probably say assumptions. The assumption is that you can bring into register the stages of morphological development—which may perhaps be slightly shifted when you compare them across species—with the gene expression data, or perhaps even literally synchronize these steps.

Christoph Hoffmann A kind of layering results: With every analysis step, the work with the data is riddled with more assumptions. But you remove the layers again and say: It didn't work here, let's go back a step. You go off in one direction and perhaps give up, start off again in another direction ...

Hans Hofmann ... or you go in multiple directions at the same time and then compare at the end to see what best explains the phenomena you observe.

Christoph Hoffmann The left column of the poster (supplement 1) proceeds from an observation and, accordingly, from a hypothesis, the hourglass. In the center column we see how you and your group attempt to pull this hypothesis into the data analysis. Of course, we see only the result. Behind this were probably a great number of attempts that were not significant, or not suitable to explain the initial observation, or to further develop the hypothesis.

Hans Hofmann Above all, you do not see how we arrived at these expression profiles, these datasets, such that they are actually comparable with each other. That's about six months' work. In fact, it's a side project that has by now taken on a life of its own. The undergrads learn from it a great deal of what is essential in data handling, analysis, statistics, etc. The large analyses are all performed at TACC, with *Stampede*, the supercomputer. For this one needs Unix and shell scripting, etc. Many of the programs that analyze the transcriptomes are written in Python, and the statistical analyses are all done in R. All of this is open source software. The Markov network brains are implemented in the programming language C++. This is done by a postdoc in the lab we collaborate with.

Christoph Hoffmann Can you tell us a bit more about the temporal framework of the project?

Times of Experimentation

Hans Hofmann We started about 20 months ago. First you have to read the literature. Then you must identify and find all of the datasets, and sometimes communicate with the original authors of the datasets to clarify questions. For instance, it may be that they only uploaded data that have already been processed in a certain way, while we want to have the original data. People have been quite helpful; we were able to get all of the data. Once we had the microarray data and RNAseq data, we had to make them compatible. There are a couple of approaches to this in the literature, with which nobody is satisfied. Therefore we invented our own approach, which took quite a while, probably longer than anything else so far. In the meantime, the colleagues from Michigan State began with the Markov network brains. Initially, they also had some problems with the question of how they should implement this without a predetermined end result, so that evolution can actually take place. After all, you don't know what's going to happen. Another question was which metrics have to be read out of the data in order to make their results comparable with our results. We're still working on that. The analyses shown on the poster were probably executed within two weeks—that went really fast. Now we have a pretty clear idea of what we still have to do at our end. Once we've taken care of that, I think we can write a first paper to lay a foundation. And then the next paper, in which we actually test the hypothesis.

Christoph Hoffmann Hans-Jörg, when you think about your lab work, would you say that the division of time has radically changed? Hans says that the preparation of the raw material, as I would call it, took most of the time.

Hans-Jörg Rheinberger When you factor in everything ... For you constantly have to replenish your supplies; components you need for the experiment have to be made available over and again; old batches have to be compared with new ones, and so on and so forth. I would say that 80 percent of the time is for experimental infrastructure work and 20 percent of the time you can do clever experiments. That's the ratio, more or less. This can always vary a bit from one area to the next.

Hans Hofmann We have lab experiments that take about the same time: one, two years easily.

Hans-Jörg Rheinberger For a substantial paper, a year or two of work is not much.

Christoph Hoffmann What's important to me is to realize how much time goes into the preparation, and now, for Hans, into the processing of the data material.

Hans Hofmann That is pretty much the lion's share.

Hannes Rickli So how does one differentiate between preparation of the infrastructure and the short experimental period? Hans, how do you do this: Is matching the data an infrastructure project?

Hans Hofmann Yes, that is a prerequisite for being able to perform the analyses at all.

Hannes Rickli So the experimental part begins with the data analysis?

Hans Hofmann That's the part where you ask the questions.

Hans-Jörg Rheinberger Let's say you are planning an experiment. You have a good idea of what you want to do and what you need. If you convert that to days of the week, I would say, you prepare it from Monday through Thursday and on Friday you get to work. A great deal of preparatory routine is involved.

Hans Hofmann Yet, with the data we massaged or otherwise prepared, we can still perform all kinds of other possible studies.

Christoph Hoffmann But what is essential is that the data from the repositories were adapted for your research. This is now part of the data. This is no big deal, of course, it is simply necessary for the task.

Hans Hofmann But it is not trivial.

Christoph Hoffmann No, it is not trivial. It shows that data are not simply available, but have to be elaborately prepared in each case.

Hans Hofmann That cannot be seen at all on the poster.

Gabriele Gramelsberger You create a data organism, like a model organism, and on this basis you can then run your experiments.

Hans Hofmann I could show you a different poster (supplement 2), which shows how this storage takes place. We even also considered whether we should publish a methods paper. We probably won't do that, but will simply bring the method into a different paper. It is relevant, but not such a major advancement.

Christoph Hoffmann That is a very interesting point, at which our different perspectives become clear. For you it is a methodological problem that has to be solved in order to get valid results.

Hans Hofmann First of all, in order to do what I want to do.

Christoph Hoffmann And for me this step itself is the focus, because it makes us understand that datasets are initially just a chaotic collection.

Hans Hofmann You are interested in how these data are transformed: The hidden steps on the poster.

Christoph Hoffmann They are not actually hidden, for you take them in full public view. They're only missing on the poster, because they do not contribute directly to the argument.

Hans Hofmann On the poster they would detract from what is essential.

Hannes Rickli I certainly find that interesting. In my early studies on the videograms we did indeed work out in detail how the media, the organisms, the behavior of the team, the camera, the environment, the light conditions, etc., configure each other; these 100 factors that register in the image; all of the preparations, the calibration—in order to then pose specific questions. Here all of that takes place completely in the digital realm. Nevertheless, even more components are probably massaged with each other, aren't they?

Hans Hofmann Science is ninety percent sweat and ten percent luck. Or you need intuition.

Gabriele Gramelsberger But you do have your own hypothesis to explain the hourglass?

Hans Hofmann Nothing really solid. We have a few ideas. The poster that we were just talking about (supplement 2), is relatively simple; an undergraduate poster for an undergraduate research forum. It explains what originally motivated the search: the hourglass, the alternative hypotheses, random variation, Haeckel's funnel. Then the poster shows the various datasets with which we are working: some that were obtained with RNAseq and some with Microarray. The different technologies are even explained briefly. I would never do this normally, but for an undergraduate poster it's ideal, because that way the student can show that she understands what she's talking about. Next the problem is rendered quantitatively. The data that are illustrated with these lines apparently do not fit each other; that has to be changed somehow. Shown here is the correction we apply and what comes out at the end. This is the pipeline strategy, according to which the data are filtered and corrected; a relatively simple method. The math behind this does not constitute any special challenge. But you do have to convince your colleagues that it is actually a valid method.

Christoph Hoffmann Does every individual dataset pass through the pipeline, or can I bundle, for instance, ten datasets that were generated with the same technology in the same laboratory?

Hans Hofmann One could do that, but we do not. There is another approach in which something quite similar happens. Let's say you have a multidimensional space and the datasets can be anywhere in this space, and you have to bring them together. Then it may make sense that the correction and filtering you do with one dataset are informed by the other datasets. This is also quite a common approach. In our case, however, it doesn't make any difference whether we do it individually or bundled. It doesn't make any difference, but it could make a difference. Some of these figures will probably later appear in the paper as supplementary material in order to explain the method. In the end, this may be just a figure with two panels. So, relatively simple.

Hannes Rickli What do the data you downloaded look like?

Hans Hofmann They are just really large text files. Different lines and columns that represent different genes, and every gene has different properties, that is, expression level, various quality flags, etc. Here on the computer screen you see such a file for the mouse, embryonic stages 10 and 10.5. This is how it comes from the repositories.

Hannes Rickli And when you plug it into an evaluation program, does it turn into rational text?

Hans Hofmann No, it will probably never produce text, but it processes the data with various analysis scripts, and ultimately you transform the result into figures.

Hannes Rickli This is probably not allowed, but I would be interested in an excerpt from the dataset you just showed us.

Hans Hofmann Of course it's allowed, since the datasets are publicly available. You can download it yourself.

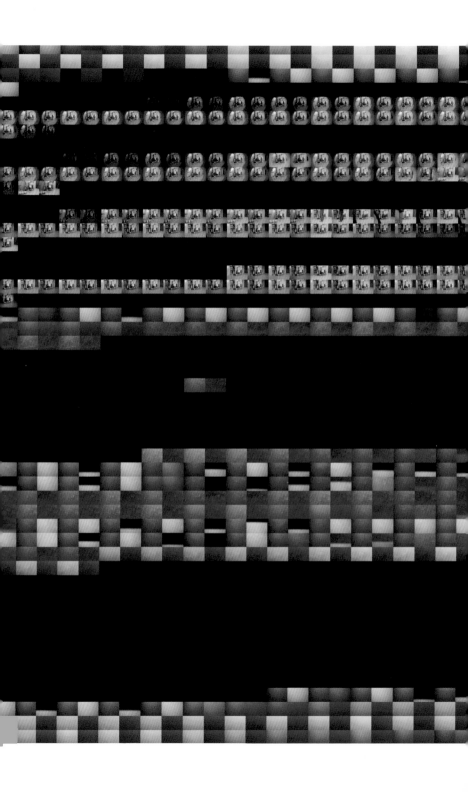

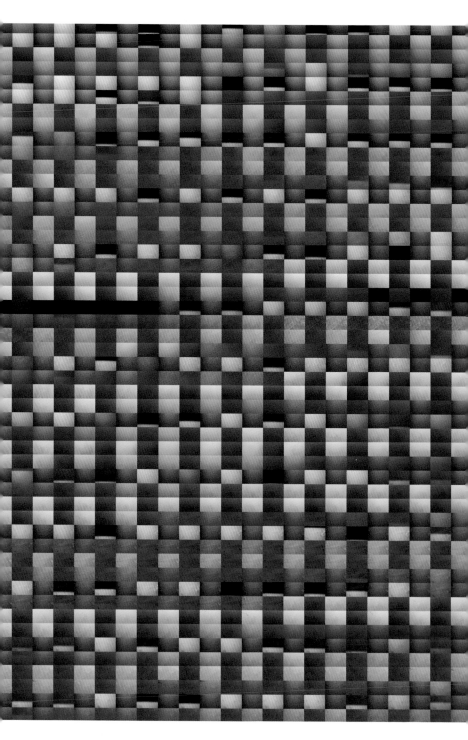

Hans-Jörg Rheinberger

Experiments, Traces, Data Streams

A Reminiscence

The conversations presented here have their origins, at least from my personal perspective, far back in the past. It all started with a "tracking sphere," a "writing game" Hannes Rickli designed back in the early 1990s. It was based on the model of a globe whose sphere registered the tiny movements of insects that came into contact with and began to crawl on it. We met for the first time in 1995 on the occasion of the *Tracking Sphere* exhibition at Kunst-Werke Berlin.[1] I was fascinated by the world of light trails that this object generated, which also integrated visitors to the exhibition room through light barriers. Rickli had read my essay *Experiment, Differenz, Schrift* and thought that these considerations on scientific experimentation could inspire his own artistic work.[2] Although this was the end of the discussion at the time, we did not lose track of each other entirely. One of my fond memories is a seminar at the Zurich University of the Arts, to which Jörg Huber and Hannes Rickli invited me during my residence at Collegium Helveticum in Zurich in the summer semester 2000. The seminar focused on Jacques Derrida's *Of Grammatology*,[3] and again, concerned the concept of the trace, which, as is well known, stands at the center of this treatise.

---

1   Hannes Rickli, *Spurenkugel – ein Schreibspiel* (Baden: Lars Müller, 1996).
2   Hans-Jörg Rheinberger, *Experiment, Differenz, Schrift: Zur Geschichte epistemischer Dinge* (Marburg/Lahn: Basiliken-Presse, 1992).
3   Jacques Derrida, *Of Grammatology*, trans. Gayatri Chakravorty Spivak (Baltimore: Johns Hopkins University Press, 2016).

When Hannes Rickli launched his project *Surplus: Video-grams of Experimentation* in 2007, a continuous exchange began to develop, which brought together artists, art historians, scientists, historians of science, as well as sociologists and philosophers interested in the sciences. Rickli had already established contacts with scientific laboratories starting in the early 1990s, among them with Hans Hofmann, who was conducting behavioral genetics studies on insects and cichlids in Leipzig at the time, and then later in Austin, Texas; with Steven Fry, who was investigating flight control by the fruit fly *Drosophila melanogaster* in Zurich; and with Philipp Fischer, who initially was studying the acoustic communication behavior of burbots in Konstanz, and then continued his studies on gurnards at the Alfred Wegener Institute on Heligoland. At that time, we began meeting at regular intervals, usually at one of the participating research facilities. Rickli tapped into the experimental data streams that were continuously generated in these laboratories, still before the filtered and cleaned products of these recorded traces were processed in scientific publications, and used them as source material for videograms that traced the gestures of not only the research objects, but also of the researchers who endeavored to understand their behavior.[4]

Surplus was the keyword. It stood for redundancy, as well as for contamination and rejects in the process of the experimental generation of traces, for the unavoidable noise still present in its fixed and processed form, the data stream. If what fascinated Rickli initially was the sheer abundance of data before they were thinned out scientifically, the exuberance before representation, this densely populated space of the graphematic, over the course of time his interest shifted more and more to the sounds and traces that result from the media of trace generation and processing themselves. He con-

---

4   Hannes Rickli, ed., *Videograms: The Pictorial Worlds of Biological Experimentation* (Zurich: Scheidegger & Spiess, 2011).

tinued following this thread in 2012 in a subsequent project that picked up on its predecessor: *Computer Signals: Art and Biology in the Age of Their Digital Experimentation,* which is now in its second funding phase. The materiality of the digital now shifted to the foreground, increasingly determining the artist's explorations and resulting in combined video and sound installations like *Fischen lauschen (Beginning of Data Transmission from the Arctic Sea)* in the project space of the Schering Stiftung in 2013. This exhibition addressed the recalcitrance of the devices that Philipp Fischer deployed at the Alfred Wegener Institute on Heligoland in order to survey the marine life at the bottom of the Arctic Sea by Spitsbergen. The ongoing project focuses on Hans Hofmann's laboratory at the University of Texas at Austin. A sound panorama leading all the way into the oilfield of the university makes perceptible to the senses the infrastructure that is needed to cool the supercomputers on the campus and keep them running. Without these energy-guzzling digital machines, such activities as the genome-based behavioral research pursued there by Hans Hofmann would be inconceivable.

In discussions among the participants that accompanied the project, the concept of data with all its facets, its embedding in computer software and infrastructure, shifted ever more to the foreground. The actual enthusiasm for the digital is counterbalanced by the massive materiality of this basis. My contributions to this conversation testify to the fact that I myself am rather skeptical about the contemporary digital data optimism. I grew up in a scientific environment in which the generation of data for the sake of data themselves had no value. That was an image linked with the scientific positivism of the 19th century. My everyday experimentation around 1980 consisted of processing experimental traces into data, which were directly or at least indirectly connected with a scientific question. The quality of these data had to be checked *ad hoc* every time in order to decide whether the question could be pursued further, or whether it had to be changed. This called for intelligent experiments, particularly ones that broke new ground. Data were gathered accordingly. The new positivism of data generation in recent decades, by contrast, aims primarily at overabundance, and is unthinkable without the availability of electronic data processing.

Yet, it cannot be overlooked that the development of suitable data repositories not only represents a technical challenge, but goes along with the emergence of new epistemological provocations.[5] In addition, it is obvious that the massive expansion of the electronic data space is leading to the establishment of more and more virtual environments, and thus to knowledge spaces in which kinds of second-order experiments are designed and performed. One example was referred to extensively in our discussion. If epistemology has a task today, it is, first and foremost, to understand the conditions of such digital experimentation at all, and to relate it to analog hard and soft, dry and wet experimentation, which no doubt will persist. Today, however, the epistemological engagement with the contemporary sciences is still a very long way from achieving this.

The two-day conversation between Philipp Fischer, Gabriele Gramelsberger, Christoph Hoffmann, Hans Hofmann, myself and Hannes Rickli in September 2016, which took place in the seclusion of Rigi Kulm in central Switzerland, was the peak of our attempts at understanding the contemporary data concept and its contexts so far; ultimately, much of our talk revolved around the above questions, both affirmatively and critically.

---

5   Sabina Leonelli, *Data-Centric Biology: A Philosophical Study* (Chicago: University of Chicago Press, 2016).

Participants

Philipp Fischer is head of the *Center for Scientific Diving* and the working group *In situ ecology and technology* at the Alfred Wegener Institute, Helmholtz Centre for Polar and Marine Research, Bremerhaven/Heligoland and Professor of Marine Biology at the Jacobs University at Bremen. His research focus is on fish behavior, fish acoustics, underwater observatories and scientific diving. Selected publications: Markus Brand and Philipp Fischer, "Species Composition and Abundance of the Shallow Water Fish Community of Kongsfjorden, Svalbard," *Polar Biology* 39 (2016): pp. 2155–2167; Philipp Fischer, Max Schwanitz, Reiner Loth, Uwe Posner, Markus Brand, and Friedhelm Schröder, "First Year of Practical Experiences of the New Arctic AWIPEV-COSYNA Cabled Underwater Observatory in Kongsfjorden, Spitsbergen," *Ocean Science* 13 (2017): pp. 259–272; Justin J. H. Buck, Scott J. Bainbridge, Eugene F. Burger, Alexandra C. Kraberg, Matthew Casari, Kenneth S. Casey, Louise Darroch, Joaquin Del Rio, Katja Metfies, Eric Delory, Philipp F. Fischer, Thomas Gardner, Ryan Heffernan, Simon Jirka, Alexandra Kokkinaki, Martina Loebl, Pier Luigi Buttigieg, Jay S. Pearlman, and Ingo Schewe, "Ocean Data Product Integration Through Innovation: The Next Level of Data Interoperability," *Frontiers in Marine Science* 6 (2019): article 32.

Gabriele Gramelsberger is Professor for Theory of Science and Technology at the RWTH Aachen University. Her current research focus is on the digitalization of science and research as well as on the machine epistemology of artificial intelligence. Selected publications: Gabriele Gramelsberger, ed., *From Science to Computational Sciences: Studies in the History of Computing and its Influence on Today's Sciences* (Zurich: diaphanes, 2011); Gabriele Gramelsberger, "Climate and Simulation," *Oxford University Research Encyclopedia Climate Science* (2018); Matthias Heymann, Gabriele Gramelsberger, and Martin Mahony, eds., *Cultures of Prediction in Atmospheric and Climate Science: Epistemic and Cultural Shifts in Computer-based Modelling and Simulation* (London: Routledge, 2019).

Christoph Hoffmann is Professor of Science Studies at the University of Lucerne. His current research focus is on data work in biology and the formation of epistemological concepts and values in academic training. Selected publications: Christoph Hoffmann, *Die Arbeit der Wissenschaften* (Zurich: diaphanes, 2013); Christoph Hoffmann, "Does a Glow Worm See? Sigmund Exner's Study of the Compound Eye," *Representations* 138 (Spring 2017): pp. 37–58; Christoph Hoffmann, *Schreiben im Forschen: Szenen Verfahren, Effekte* (Tübingen: Mohr Siebeck, 2018); Michael Hagner and Christoph Hoffmann, eds., *Materialgeschichten*, Nach Feierabend: Zürcher Jahrbuch für Wissensgeschichte, vol. 14 (Zurich: diaphanes, 2018).

Hans Hofmann is Professor of Integrative Biology at The University of Texas at Austin. He is an evolutionary neuroscientist, who uses genomic approaches to uncover the neural and molecular underpinnings of social evolution. He has developed and led several successful training programs in systems neuroscience, computational biology, and bioinformatics. Selected publications: Chelsea A. Weitekamp and Hans A. Hofmann, "Evolutionary Themes in the Neurobiology of Social Cognition," *Current Opinion in Neurobiology* 28 (2014): pp. 22–27; Hans A. Hofmann, Annaliese K. Beery, Daniel T. Blumstein, Iain D. Couzin, Ryan L. Earley, Loren D. Hayes, Peter L. Hurd, Eileen A. Lacey, Steven M. Phelps, Nancy G. Solomon, Michael Taborsky, Larry J. Young, and Dustin R. Rubenstein, "An Evolutionary Framework for Studying Mechanisms of Social Behavior," *Trends in Ecology & Evolution* 29 (2014): pp. 581–589; Dustin R. Rubenstein, J. Arvid Ågren, Lucia Carbone, Nels C. Elde, Hopi E. Hoekstra, Karen M. Kapheim, Laurent Keller, Corrie S. Moreau, Amy L. Toth, Sam Yeaman, and Hans A. Hofmann, "Coevolution of Genome Architecture and Social Behavior," *Trends in Ecology & Evolution* 34 (2019): pp. 844–855.

Hans-Jörg Rheinberger is Director emeritus at the Max Planck Institute for the History of Science in Berlin. His research interests revolve around the practices of experimentation in the sciences and in the arts. Selected publications: Hans-Jörg Rheinberger and Staffan Müller-Wille, *The Gene: From Genetics to Postgenomics* (Chicago: University of Chicago Press,

2017); Hans-Jörg Rheinberger, *The Hand of the Engraver: Albert Flocon Meets Gaston Bachelard* (Albany: State University of New York Press, 2018); Hans-Jörg Rheinberger, *Experimentalität* (Berlin: Kadmos Verlag, 2018).

Hannes Rickli is a visual artist. He teaches and researches as a Professor at the Zurich University of the Arts. His main focuses are the materiality of the digital, the instrumental use of media and space as well as media ecology. Selected publications: Hannes Rickli, ed., *Videograms: The Pictorial Worlds of Biological Experimentation as an Object of Art and Theory* (Zurich: Scheidegger & Spiess, 2011); Hannes Rickli, "Experimentieren," in *Künstlerische Forschung: Ein Handbuch*, ed. Jens Badura et al. (Zurich: diaphanes, 2015); Hannes Rickli and Valentina Vuksic, *RemOs1: Beginning Data Work in the Arctic Sea* (Zurich: Zurich University of the Arts, 2016); Hannes Rickli, "Der unsichtbare Faden: Zu Materialität und Infrastrukturen digitaler Tierbeobachtung," *Zeitschrift für Medien- und Kulturforschung ZMK* 2016, no. 2 (2016): pp. 137–154.

List of Illustrations

Cover — *RemOs1*, excerpt from the archive of stereometric images, Kongsfjord, Ny-Ålesund, Spitsbergen, 2012–2014. Thumbnails of image pairs, May 11–24, 2013. © Philipp Fischer, Hannes Rickli

PP. 7-8 — Fig. 1, 2: *RemOs1*, archive of stereometric images, Kongsfjord, Ny-Ålesund, Spitsbergen. Thumbnails of image pairs, April 3, 2013, 06:00:43 to April 28, 2013, 15:01:01. © Philipp Fischer, Hannes Rickli

PP. 29-30 — Fig. 3, 4: Evaluation computer for stereometric images *RemOs1*, April 24, 2014. © Hannes Rickli

PP. 63-64 — Fig. 5, 6: Test protocol and evaluation of codon–anticodon interactions in ribosomal E sites, February 1985. © Hans-Jörg Rheinberger

P. 95 — Fig. 7: *RemOs1*, Kongsfjorden, Ny-Ålesund, Spitsbergen, March 9, 2013, 15:30:12. © Philipp Fischer

P. 96 — Fig. 8: *RemOs1*, Kongsfjorden, Ny-Ålesund, Spitsbergen, June 24, 2013, 13:58:20. © Philipp Fischer

P. 115 — Fig. 9: *Cichlid #3*, location 5: Supercomputer Stampede, Texas Advanced Computing Center (TACC), Research Office Complex (ROC), J.J. Pickle Research Campus UT Austin. Average ambient temperature 57.4 °F/14 °C. Coordinates N 30°23'24.9", W 97°43'30.8". © Birk Weiberg, Hannes Rickli

P. 116 — Fig. 10: *Cichlid #3*, location 8: Rig #641, fracking drill. University Lands UT Texas, Crane County. Average ambient temperature 95.0 °F/35 °C. Coordinates N 31°31'44.6", W 102°26'55.8". © Birk Weiberg, Hannes Rickli

PP. 139-140 — Fig. 11, 12: *RemOs1*, archive of stereometric images, Kongsfjord, Ny-Ålesund, Spitsbergen. Thumbnails of image pairs, March 15, 2014, 05:30:29 to April 9, 2014, 09:30:28. © Philipp Fischer, Hannes Rickli

Index

VOLUME 22 IN THE SERIES OF THE

INSTITUTE FOR CONTEMPORARY ART RESEARCH (IFCAR),

ZURICH UNIVERSITY OF THE ARTS (ZHDK)

FIRST PRINTING

ISBN 978-3-0358-0224-5

© DIAPHANES, ZURICH 2020

COVER IMAGE: *REMOS1*, EXCERPT FROM THE ARCHIVE OF STEREOMETRIC IMAGES,

KONGSFJORD, NY-ÅLESUND, SPITSBERGEN, 2012–2014. THUMBNAILS OF

IMAGE PAIRS, MAY 11–24, 2013. © PHILIPP FISCHER, HANNES RICKLI

LAYOUT: 2EDIT, ZÜRICH

PRINTED IN GERMANY

WWW.DIAPHANES.COM